Texts in Computer Science

Editors
David Gries
Fred B. Schneider

For further volumes:
http://www.springer.com/series/3191

Sholom M. Weiss • Nitin Indurkhya • Tong Zhang

Fundamentals of Predictive Text Mining

 Springer

Sholom M. Weiss
T.J. Watson Research Center
IBM Corporation
Kitchawan Road 1101
Yorktown Heights, 10598 NY
USA
sholom@data-miner.com

Nitin Indurkhya
School of Computer Science & Engg.
University of New South Wales
Sydney, 2052 NSW
Australia
nitin@data-miner.com

Tong Zhang
Dept. Statistics, Hill Center
Rutgers University
Piscataway, 08854-8019 NJ
USA
tongz@rci.rutgers.edu

Series Editors
David Gries
Department of Computer Science
Upson Hall
Cornell University
Ithaca, NY 14853-7501, USA

Fred B. Schneider
Department of Computer Science
Upson Hall
Cornell University
Ithaca, NY 14853-7501, USA

ISBN 978-1-4471-2565-5
DOI 10.1007/978-1-84996-226-1
Springer London Dordrecht Heidelberg New York

ISBN 978-1-84996-226-1(eBook)

British Library Cataloguing in Publication Data
A catalogue record for this book is available from the British Library

Preface

Five years ago, we authored "Text Mining: Predictive Methods for Analyzing Unstructured Information." That book was geared mostly to professional practitioners, but was adaptable to course work with some effort by the instructor. Many topics were evolving, and this was one of the earliest efforts to collect material for predictive text mining. Since then, the book has seen extensive use in education, by ourselves and other instructors, with positive responses from students. With more data sourced from the Internet, the field has seen very rapid growth with many new techniques that would interest practitioners. Given the amount of supplementary new material we had begun using, a new edition was clearly needed. A year ago, our publisher asked us to update the book and to add material that would extend its use as a textbook. We have revised many sections, adding new material to reflect the increased use of the web. Exercises and summaries are also provided.

The prediction problem, looking for predictive patterns in data, has been widely studied. Strong methods are available to the practitioner. These methods process structured numerical information, where uniform measurements are taken over a sample of data. Text is often described as unstructured information. So, it would seem, text and numerical data are different, requiring different methods. Or are they? In our view, a prediction problem can be solved by the same methods, whether the data are structured numerical measurements or unstructured text. Text and documents can be transformed into measured values, such as the presence or absence of words, and the same methods that have proven successful for predictive data mining can be applied to text. Yet, there are key differences. Evaluation techniques must be adapted to the chronological order of publication and to alternative measures of error. Because the data are documents, more specialized analytical methods may be preferred for text. Moreover, the methods must be modified to accommodate very high dimensions: tens of thousands of words and documents. Still, the central themes are similar.

Our view of text mining allows us to unify the concepts of different fields. No longer is "natural language processing" the sole domain of linguists and their allied computer specialists. No longer is search engine technology distinct from other forms of machine learning. Ours is an open view. We welcome you to try your hand at learning from data, whether numerical or text. Large text collections, often readily available on the Internet, contain valuable information that can be mined with today's tools instead of waiting for tomorrow's linguistic techniques. While others search for the essence of language understanding, we can immediately look for recurring word patterns in large collections of digital documents.

Our main theme is a strictly empirical view of text mining and an application of well-known analytical methods. We provide examples and software. Our presentation has a pragmatic bent with numerous references in the research literature for you to follow when so inclined. We want to be practical, yet inclusive of the wide community that might be interested in applications of text mining. We concentrate on predictive learning methods but also look at information retrieval and search engines, as well as clustering methods. We illustrate by examples, case studies, and the accompanying downloadable software.

While some analytical methods may be highly developed, predictive text mining is an emerging area of application. We have tried to summarize our experiences and provide the tools and techniques for your own experiments.

Audience

Our book is aimed at IT professionals and managers as well as advanced undergraduate computer science students and beginning graduate students. Some background in data mining is beneficial but is not essential. A few sections discuss advanced concepts that require mathematical maturity for a proper understanding. In such sections, intuitive explanations are also provided that may suffice for the less advanced reader. Most parts of the book can be read and understood by anyone with a sufficient analytic bend. If you are looking to do research in the area, the material in this book will provide direction in expanding your horizons. If you want to be a practitioner of text mining, you can read about our recommended methods and our descriptions of case studies. The software requires familiarity with running command-line programs and editing configuration files.

For Instructors

The material in this book has been successfully used for education in a variety of ways ranging from short intensive one-week courses to twelve-week full semester courses. In short courses, the mathematical material can be skipped. The exercises have undergone class-testing over several years. Each chapter has the following accompanying material:

- a chapter summary
- exercises.

In addition, numerous examples and figures are interlaced throughout the book, and these are available at data-miner.com as freely downloadable slides.

Supplementary Web Software

Data-Miner Pty. Ltd. has provided a free software license for those who have purchased the book. The software, which implements many of the methods discussed in the book, can be downloaded from the data-miner.com Web site. Linux scripts for many examples are also available for download. See the Appendix A for details.

Acknowledgements

Fred Damerau, our colleague and mentor, was a co-author of our original book. He is no longer with us, and his contributions to our project, especially his expertise in linguistics, were immeasurable. Some of the case studies in Chap. 7 are based on our prior publications. In those projects, we acknowledge the participation of Chidanand Apté, Radu Florian, Abraham Ittycheriah, Vijay Iyengar, Hongyan Jing, David Johnson, Frank Oles, Naval Verma, and Brian White. Arindam Banerjee made many helpful comments on a draft of our book. The exercises in the book evolved from our text-mining course conducted regularly at statistics.com. We thank our editor, Wayne Wheeler, and our previous editors Ann Kostant and Wayne Yuhasz, for their support.

New York, USA
Sydney, Australia

Sholom Weiss and Tong Zhang
Nitin Indurkhya

Contents

Chapter 1
Overview of Text Mining

1.1 What's Special About Text Mining?

Do you have a shortage of data? Not very likely. A consequence of the pervasive use of computers is that most data originate in digital form. If we trade a stock or write a book or buy a product online, these events evolve electronically. Since so many paper transactions are now in paperless digital form, lots of "big" data are available for further analysis.

The concept of data mining, finding valuable patterns in data, is an obvious response to the collection and storage of large volumes of data. Data mining is no longer an emerging technology awaiting further development. Although its application is far from universal, the techniques of data mining are highly developed and for some forms of analysis are entering a mature phase.

We would like to say "Give us data and we will find the patterns." Unfortunately, data-mining methods expect a highly structured format for data, necessitating extensive data preparation. Either we have to transform the original data, or the data are supplied in a highly structured format.

Data-mining methods learn from samples of past experience. If we speak to specialists in predictive data mining, their data will be in numerical form. These people are the "numbers guys." The "text miners" do not expect an orderly series of numbers. They are happy to look at collections of documents, where the contents are readable and their meaning is obvious.

This is our first distinction between data and text mining: numbers vs. text. That doesn't mean that these are two distinct concepts. Both are based on samples of past examples. The composition of the examples is very different, yet many of the learning methods are similar. That's because the text will be processed and transformed into a numerical representation.

S.M. Weiss et al., *Fundamentals of Predictive Text Mining*,
Texts in Computer Science 41,
DOI 10.1007/978-1-84996-226-1_1, © Springer-Verlag London Limited 2010

Fig. 1.1 Structured data in standard format

Fig. 1.2 A spreadsheet example of medical data

Gender	Systolic BP	Weight	Disease Code
M	175	65	3
F	141	72	1
...	...	...	...
F	160	59	2

1.1.1 Structured or Unstructured Data?

Superficially, we see numbers or text in our data. The text is usually a collection of unstructured documents with no special requirements for composing the documents. As noted above, most data-mining applications assimilate only structured information. The data must be prepared in a very special way before any learning methods can be applied. Figures 1.1 and 1.2 illustrate the world of structured data.

Typical data-mining applications use structured information that is carefully prepared. The data may be transformed by a "data preparation" process or, better yet, the data may be collected based on careful prior design for mining. The items that will be used are clearly described over a range of all possibilities, and these are then recorded uniformly for every example that is a member of the sample. The recipe is well-known. Two types of information are expected: (a) ordered numerical and (b) categorical. Ordered numerical attributes have values where greater than or less than comparisons have meaning. For example, weight and income are obviously ordered. Categorical attributes are unordered numerical codes that have a definition in a codebook. The most common categorical attribute is something that can be measured as true or false, represented by a one or a zero. For example, gender can be measured as male or female, or business category can be measured by a code. The meaning of the code is described elsewhere, not to be used by the learning program but by the individuals interpreting the results of learning.

If data can be described by a spreadsheet with its tabular format, then the problem is highly structured. The task of data collection is to fill in the blank cells. Many mathematical methods use spreadsheet data, also known as a matrix. To learn from spreadsheet data, we populate the cells formed by intersecting rows and columns. We must fill in these cells in a uniform manner. Each cell is organized in the same way. A completed row is a complete example of past experience. For example, in a medical domain, it might be a single patient. A column is an example of one measurement on the patient, for example blood pressure. Thus, the intersection of row 2 and column 3 in Fig. 1.2 is the weight of the second patient or, more generally, the second example and third measurement.

We can now clearly see the structured world of data mining. Data must be represented in a highly organized manner. We typically use a spreadsheet format. The labels of columns are designed to fit the domain and are made permanent. Following this design, data are collected by adding rows (i.e., examples), where each example is measured using the same attributes. It is mechanically easy to add a new example by filling in a row. Adding a new attribute, a column, is more difficult, requiring a review of all previous examples and a measurement of the new value for each. Once we have data such as these, we can operate in a typical mathematical fashion. A single row or column is a vector, and the complete spreadsheet or database table is a matrix.

1.1.2 Is Text Different from Numbers?

The presentations of data for classical data mining and text mining are quite different. Whereas data-mining methods like to see the data in spreadsheet format, text-mining methods like to see a document format, and the standard presentation for learning is a variant of the format called XML used in the document world. Clearly, we expect that text is quite different from numbers. Still, the methods that we will discuss in this book are similar to those used for data mining. These methods have proved remarkably successful without understanding specific properties of text such as the concepts of grammar or the meaning of words. Strictly low-level frequency information is used, such as the number of times a word appears in a document, and then well-known methods of machine learning are applied. One of the main themes supporting text mining is the transformation of text into numerical data, so although the initial presentation is different, at some intermediate stage, the data move into a classical data-mining encoding. The unstructured data become structured.

Our text-mining methods will be similar to classical data-mining methods. These methods will transform data from text to standard numerical forms. To make these methods work, we need to transform text into a standard spreadsheet format and fill in the spreadsheet's cells. The rows of a spreadsheet are examples of prior experience, so for text, we can consider a document to be one complete example. A column is an attribute that can be measured. In the most fundamental model of text, we can consider the presence or absence of a word to be a measured attribute for each document. Thus, each row represents a document and each column a word. We could fill in the cells, as in Fig. 1.3, with ones or zeros. In this example, the word "income" appears in documents 1 and 3 but not documents 2 or 4.

Company	Income	Job	Overseas
0	1	0	1
1	0	1	1
1	1	1	0
0	0	0	1

Fig. 1.3 A binary spreadsheet of words in documents

Many variants of this document and word representation could be explored, but this is the fundamental concept, where words are attributes and documents are examples, and together these form a sample of data that can feed our well-known learning methods. Many machine-learning methods perform accurately with this transformation, working with far larger amounts of data than humans could hope to process. These programs have little knowledge of meaning or grammar. They are statistical methods that lack prior knowledge. They counterbalance that deficiency with massive processing of data, finding patterns in word combinations that are recurring and predictive.

The spreadsheet model of data returns us to the familiar territory of classical data-mining methods. Nevertheless, we would be foolish to rush to apply learning methods in their original form without taking advantage of the special characteristic of text. The spreadsheet remains the conceptual model, but it would be impractical, inefficient, or even ineffective until we understood some of its important differences from classical numerical data.

Consider a collection of documents. The set of attributes will be the total set of unique words in the collection. We call this set of words a dictionary. The examples are the individual documents. We compose a spreadsheet and fill in the cells with a one for the presence of a word and a zero for its absence. An application might have many thousands or even millions of documents. The dictionary will converge to a smaller number of words than the number of documents but can readily number several hundred thousands. Specialized documents, such as repair manuals with part numbers that are alphanumeric, may lead to very large dictionaries. It appears that the spreadsheet model is too unwieldy to be practical.

Viewing the spreadsheet more closely, we see almost all zeros. Unless individual documents are surprisingly lengthy, almost book length, the matrix is sparse: any individual document will use only a tiny subset of the potential set of words in a dictionary. Because of that special characteristic, the spreadsheet remains a reasonable conceptual model of data. Methods that process text will expect sparse spreadsheets and will leverage that property in their implementations to store only positive cell values.

Sparseness is not the only representational difference. All the values in a text-mining spreadsheet are positive. Classical data-mining methods will consider all values of an attribute, both positive and negative. The decision criteria could readily say "if word x has value zero, then conclude class y." In contrast, text-mining methods mostly concentrate on positive matches, not worrying whether other words are absent from a document. This view also leads to great simplifications in processing, often allowing text-mining programs to operate in what would be considered huge dimensions for regular data-mining applications.

If we focus on positive occurrences of words, we also have a solution to one of the *bête noires* of applying data-mining methods: missing values. The spreadsheet model for data has a cell for each measurable value in an example. Most methods expect the cell to have a value. In practical applications, such as when we extract information from a real-world database, a great deal of information is missing, and the cell remains empty. An empty cell is not the same as saying that the answer is a default value, such as false for a binary-valued attribute or a mean value for a real-valued attribute. Many schemes have been developed for managing missing values, almost all with inherent deficiencies. These weaknesses are particularly manifest when the majority of values are missing. For text, missing values are a nonissue: words are either present or absent from a document. We can completely fill in the spreadsheet and all the cells.

In our simplified world of text mining, we have described documents as examples and words as attributes in a spreadsheet. Although it could be argued that these are gross simplifications of the representation needed for text, it is consistent with our theme of transforming words to numbers, so that known data-mining methods can be applied. We will present numerous variations on this model of data and its statistical view of words and text. Thus, although text-mining operates in very high dimensions, in many situations, processing is effective and efficient because of the sparseness characteristic of most documents and most practical applications.

Let's look at the types of problems that we can try to solve with this approach to data representation and learning methods.

1.2 What Types of Problems Can Be Solved?

A primary focus of our attention is classification and prediction. These are among the most widely studied and applied methods and applications of data mining. Given a sample of past experience and correct answers for each example, the objective is to find the correct answers for new examples. We will consider those types of problems, such as text categorization, that are clear applications for predictive methods.

The concept of classification can be extended to data that do not have clearly labeled answers. Our task would be to organize the data in such a way that we can make up labels or answers and expect these to hold in the future. This process is referred to as clustering.

Although similarity between documents is an essential ingredient in organizing unlabeled documents into distinct groups, measuring similarity of documents is an end in itself. Measuring similarity between documents is fundamental to most forms of document analysis, especially information retrieval.

The applications that we discuss do not emphasize linguistic analysis. Statistical and associational relationships are the basis of our presentation. At some point in the future, a deeper semantic understanding may demonstrate clear performance advantages. For now, the preeminence of statistical approaches has been shaped by the increasing capabilities of computer resources. Most important has been the outpouring of digital data, where libraries of documents are in digital form, ready for analysis by text-mining methods.

Let's look at some of the areas where these text-mining methods can work for us.

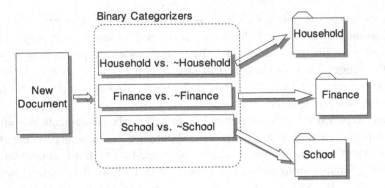

Fig. 1.4 Text categorization

1.3 Document Classification

Text categorization is the widely used, but ponderous, name for document classification. It is the purest embodiment of the spreadsheet model with labeled answers. Once the data are transformed to the usual numerical spreadsheet format, standard data-mining methods are applicable. Figure 1.4 illustrates the document classification application. Documents are organized into folders, one folder for each topic. A new document is presented, and the objective is to place this document in the appropriate folders. For example, we might have a folder for household or financial documents and we want to add new documents to the correct folder. The application is almost always binary classification because a document can usually appear in multiple folders.

Originally, this type of problem was considered a form of indexing, much like the index of a book. As more and more documents have become available online, the applicability of this task has broadened. Some of the more obvious tasks are related to e-mail: for example, automatically forwarding e-mail to the appropriate company department or detecting spam mail. The spreadsheet model with one column corresponding to the correct answer is the universal classification model for data, and the transformed text data can readily be combined with standard numerical data-mining data. As an example, you might think that you could predict future stock movements based on prior experience. You collect news articles that appeared prior to a rise or fall in stock prices along with company financial data. The labels would be binary, 1 for up and 0 for down.

1.4 Information Retrieval

Information retrieval is the topic most commonly associated with online documents. What is more fundamental to browsing the Internet than a search engine? The general task of information retrieval is illustrated in Fig. 1.5. A collection of documents is obtained, we give clues as to the documents that we want to retrieve from the

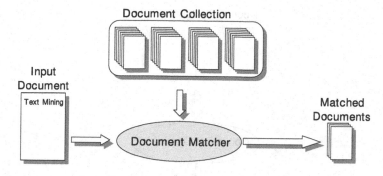

Fig. 1.5 Retrieving matched documents

collection, and then documents matching the clues are presented as answers to our query.

What are the clues, and how are they used to retrieve relevant documents? The clues are words that help identify the relevant stored documents. In a typical instance of invoking a search engine, a few words are presented, and these words are matched to the stored documents. The best matches are presented as the responses. The process can be generalized to a document matcher, where instead of a few words, a complete document is presented as a set of clues. The input document is then matched to all stored documents, retrieving the best-matched documents.

A basic concept for information retrieval is measuring similarity: a comparison is made between two documents, measuring how similar the documents are. For comparison, even a small set of words input into a search engine can be considered as a document that can be matched to others. From one perspective, measuring similarity is related to predictive methods for learning and classification that are called nearest-neighbor methods. The common theme is measuring similarity, and variations of these methods are fundamental to information retrieval.

The spreadsheet model of data can readily be used for these tasks. The new document is equivalent to a new row. The new row is compared to all the other rows, and the most similar rows and their associated documents are the answers.

1.5 Clustering and Organizing Documents

For text categorization, we saw that the objective was to place new documents into the appropriate folders. These folders were created by someone with knowledge of the document structure, someone who knew the expected topics. What if we have a collection of documents with no known structure? For example, a company may have a help desk that receives and records calls by users of their products. The company might want to learn about the categories and types of complaints. The general objective is illustrated in Fig. 1.6. Given a collection of documents, find a set of folders such that each holds similar documents.

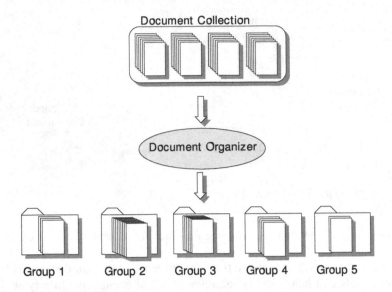

Fig. 1.6 Organizing documents into groups

The clustering process is equivalent to assigning the labels needed for text categorization. Because there are many ways to cluster documents, it is not quite as powerful a process as assigning answers (i.e., known correct labels) to documents. Still, clustering can be insightful. By studying key words that characterize a cluster, a company could learn about its customers. In the help-desk example, a computer company might find that the largest cluster of complaints is for networking problems. It might also identify an unexpected type of problem, documents indicating many calls for help, for which they do not have a good solution.

In terms of the spreadsheet model, clustering will add at least one column to a spreadsheet, corresponding to true-or-false labels for the examples. The number of labels will be determined by the clustering algorithm.

1.6 Information Extraction

Our representation of data looks at information in terms of words. This is a rudimentary formulation that is surprisingly successful for many applications. In comparison with classical numerical data-mining representations, these measurements are very shallow. We are only measuring the presence or absence of words. A data-mining representation may look the same, but the measurements themselves will be far broader and more complex. They may be real-valued variables, such as sales volume, or a code, such as an industry code like "auto industry." Someone is responsible for defining these attributes and depositing them in a database.

An alternative way of looking at these distinctions is that data mining expects highly structured data, and text is naturally unstructured. To make text structured, we

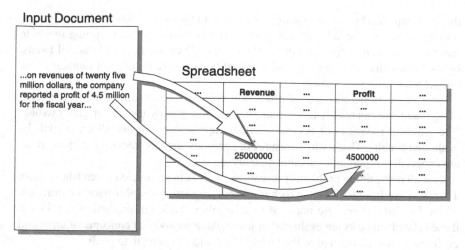

Fig. 1.7 Extracting information from a document

have employed a very shallow representation that measures the simple occurrence of words. Information extraction is a subfield of text mining that attempts to move text mining onto an equal footing with the structured world of data mining. Figure 1.7 illustrates the task of information extraction. The objective is to take an unstructured document and automatically fill in the values of a spreadsheet.

A database, at least one organized by fields or tables, is structured. When the information is unstructured, such as that found in a collection of documents, then a separate process is needed to extract data from an unstructured format. For example, we may examine documents about companies and extract the sales volumes and industry codes from text that has not been structured for storage in a database. The attribute that is measured will not have a fixed position in the text and may not be described in the same way in different documents.

In terms of our spreadsheet model of data, the objective is to fill in the cells. The role of the examples is unchanged; they are documents. The columns are not just words but can be higher-level concepts that are found by the information extraction process. This process examines documents and fills in the cells, a process that is equivalent to populating a database table. Once the process is completed, the usual learning methods can be applied to the spreadsheet.

1.7 Prediction and Evaluation

Our ultimate goal is prediction, projecting from a sample of prior examples to new unseen examples. The learning program studies documents and finds some generalized rules that will give correct answers on new examples. How do we know that we will be successful on new examples? The classic approach is to "hold out" some examples with known answers, not allowing the learning program to train on those examples. These new examples are used solely for evaluation. For many text-mining scenarios, the holdout evaluation will be effective (e.g., assigning topics to news stories, such as financial or sports stories). Even so, there are special twists. News stories change over time, and we must be sensitive to dates of publication in our selection of test documents.

One of the basic concepts of prediction is the measurement of error. For topic assignment, we can readily determine whether a program's answer is right or wrong. The classical measures of accuracy will be applicable, but not all errors will be evaluated equally. That's why measures of accuracy such as "recall" and "precision" are especially important to document analysis.

Not all text-mining problems present themselves in completed spreadsheet form. Therefore we need to examine techniques for clustering or information extraction, where labeled answers are not readily accessible. Although prediction is primary, these related subtasks are evaluated in less certain terms. The concepts of error and evaluation must be tailored to the task at hand and its immediate goals.

1.8 The Next Chapters

We will discuss each of the aforementioned topics in-depth. The emphasis is almost exclusively on a statistical approach. Although we have said that text can be mapped into a standard framework for numerical data mining, the differences in data preparation strongly influence the designs and selection of learning methods.

Having an efficient representation suitable for operation in a high-dimensional space of documents does not give a complete picture of text-mining methods. It is not just a matter of transforming words to vectors and applying standard learning methods. Experience with text and statistical methods gives us direction in favoring some approaches over others.

We will discuss our selection of methods for processing text. Document and text processing has a very large body of knowledge. Our objectives are much more limited. The presentation of text mining emphasizes predictive methods. Ours is not an encyclopedic body of knowledge. Rather, we unify several areas that have been treated separately. Our goal is a practical and introductory guide, that integrates related topics and provides practical advice for text mining.

Historically, these methods have their antecedents in several research communities, many originally associated with artificial intelligence, including the specialized fields of informational retrieval, knowledge discovery and data mining, and natural language.

1.9 Summary

Text mining and data mining are contrasted relative to automated prediction. Models are constructed by training on samples of unstructured documents, and results are projected to new text. A standard data format for input to prediction methods is described. The key objective of data preparation is to transform text into a numerical format, eventually sharing a common representation with numerical data mining. Different text-mining problems are introduced that fit within the prediction framework. These include document classification, information retrieval, clustering documents, information extraction, and performance evaluation.

1.10 Historical and Bibliographical Remarks

The roots of what we now call text mining are deep in the area of information retrieval. Document classification is similar in many ways to document indexing which was extensively studied in the late 1950s and 1960s (Luhn 1959; Maron and Kuhns 1960). Document clustering and measurement of document similarity are also old topics (Jardine and van Rijsbergen 1971). Representation of a document as a bag of words, each with an attached frequency measure, became popular by the 1970s (Salton *et al.* 1975). As artificial intelligence methods became more popular in the 1980s, they also were applied to problems we now classify as text mining. This is particularly true of text categorization (Hayes and Weinstein 1990).

The real impetus for the kinds of applications we see today comes from two sources: the availability of cheap, fast computing and of enormous amounts of text in digital form. In the past, when text collections were formed by punching them into cards or paper tape before being read into an expensive computer with limited memory, only a few centers could do the kinds of experiments now conducted on desktop computer systems.

Much of the advancement of technology in text mining has come in connection with government-sponsored challenge competitions, each capped by a conference in which the participants present their results. Examples include the MUC (Message Understanding Conference), which concluded with MUC-7, TREC (Text Retrieval Conference), CoNLL (Conference on Natural Language Learning), and ACE (Automatic Content Extraction) conferences. More general conferences of interest include the ACL (Association for Computational Linguistics) meetings, KDD (Knowledge Discovery and Data Mining), IJCAI (International Joint Conference on Artificial Intelligence), and the International Machine Learning Conference.

Although there is some overlap of interests, particularly in information extraction, for the most part our concerns and interests are quite different from those of the natural language processing community. For example, we exclude topics such as parsing (although we will discuss the utility of parsed output), dialogue understanding, deep semantic representations, and the like. In recent years, computational linguists have turned to statistics in the study of some of these problems and sometimes use the same or similar statistical methods. In particular, the speech understanding and parsing communities make extensive use of hidden Markov models.

1.11 Questions and Exercises

1. How does information extraction differ from information retrieval?
2. Give any three differences between spreadsheets for text-mining data and spreadsheets for general data-mining problems.
3. The book comes with free downloadable software. Read the Appendix A and install both tmsk and riktext.
4. Download the file ExerciseFiles.zip from the download area (see the Appendix A) and get the zipped Reuters train and test files, trn.zip and tst.zip, from the website. They are part of the Reuters-21578 data. These files will be used in the exercises in later chapters.

Chapter 2
From Textual Information to Numerical Vectors

To mine text, we first need to process it into a form that data-mining procedures can use. As mentioned in the previous chapter, this typically involves generating features in a spreadsheet format. Classical data mining looks at highly structured data. Our spreadsheet model is the embodiment of a representation that is supportive of predictive modeling. In some ways, predictive text mining is simpler and more restrictive than open-ended data mining. Because predictive mining methods are so highly developed, most time spent on data-mining projects is for data preparation. We say that text mining is unstructured because it is very far from the spreadsheet model that we need to process data for prediction. Yet, the transformation of data from text to the spreadsheet model can be highly methodical, and we have a carefully organized procedure to fill in the cells of the spreadsheet. First, of course, we have to determine the nature of the columns (i.e., the features) of the spreadsheet. Some useful features are easy to obtain (e.g., a word as it occurs in text) and some are much more difficult (e.g., the grammatical function of a word in a sentence such as subject, object, etc.). In this chapter, we will discuss how to obtain the kinds of features commonly generated from text.

2.1 Collecting Documents

Clearly, the first step in text mining is to collect the *data* (i.e., the relevant documents). In many text-mining scenarios, the relevant documents may already be given or they may be part of the problem description. For example, a Web page retrieval application for an intranet implicitly specifies the relevant documents to be the Web pages on the intranet. If the documents are readily identified, then they can be obtained, and the main issue is to cleanse the samples and ensure that they are of high quality. As with nontextual data, human intervention can compromise the integrity of the document collection process, and hence extreme care must be exercised. Sometimes, the documents may be obtained from document warehouses or databases. In these scenarios, it is reasonable to expect that data cleansing was done before deposit and we can be confident in the quality of the documents.

S.M. Weiss et al., *Fundamentals of Predictive Text Mining*,
Texts in Computer Science 41,
DOI 10.1007/978-1-84996-226-1_2, © Springer-Verlag London Limited 2010

In some applications, one may need to have a data collection process. For instance, for a Web application comprising a number of autonomous Web sites, one may deploy a software tool such as a Web crawler that collects the documents. In other applications, one may have a logging process attached to an input data stream for a length of time. For example, an e-mail audit application may log all incoming and outgoing messages at a mail server for a period of time.

Sometimes the set of documents can be extremely large and data-sampling techniques can be used to select a manageable set of relevant documents. These sampling techniques will depend on the application. For instance, documents may have a time stamp, and more recent documents may have a higher relevance. Depending on our resources, we may limit our sample to documents that are more useful.

For research and development of text-mining techniques, more generic data may be necessary. This is usually called a corpus. For the accompanying software, we mainly used the collection of Reuters news stories, referred to as Reuters corpus RCV1, obtainable from the Reuters Corporation Web site. However, there are many other corpora available that may be more appropriate for some studies.

In the early days of text processing (1950s and 1960s), one million words was considered a very large collection. This was the size of one of the first widely available collections, the Brown corpus, consisting of 500 samples of about 2000 words each of American English texts of varying genres. A European corpus, the Lancaster-Oslo-Bergen corpus (LOB), was modeled on the Brown corpus but was for British English. Both these are still available and still used. In the 1970s and 1980s, many more resources became available, some from academic initiatives and others as a result of government-sponsored research. Some widely used corpora are the Penn Tree Bank, a collection of manually parsed sentences from the Wall Street Journal; the TREC (Text Retrieval and Evaluation Conferences) collections, consisting of selections from the Wall Street Journal, the New York Times, Ziff-Davis Publications, the Federal Register, and others; the proceedings of the Canadian Parliament in parallel English–French translations, widely used in statistical machine translation research; and the Gutenberg Project, a very large collection of literary and other texts put into machine-readable form as the material comes out of copyright. A collection of Reuters news stories called Reuters-21578 Distribution 1.0 has been widely used in studying methods for text categorization.

As the importance of large text corpora became evident, a number of organizations and initiatives arose to coordinate activity and provide a distribution mechanism for corpora. Two of the main ones are the Linguistic Data Consortium (LDC) housed at the University of Pennsylvania and the International Computer Archive of Modern and Medieval English (ICAME), which resides in Bergen, Norway. Many other centers of varying size exist in academic institutions. The Text Encoding Initiative (TEI) is a standard for text collections sponsored by a number of professional societies concerned with language processing. There a number of Web sites devoted to corpus linguistics, most having links to collections, courses, software, etc.

Another resource to consider is the World Wide Web itself. Web crawlers can build collections of pages from a particular site, such as Yahoo, or on a particular topic. Given the size of the Web, collections built this way can be huge. The main problem with this approach to document collection is that the data may be of dubious quality and require extensive cleansing before use. A more focused corpus can be built from the archives of USENET news groups and accessible from many ISPs directly or through Google Groups. These discussion groups cover a single topic, such as fly fishing, or broader topics such as the cultures of particular countries. A similar set of discussions is available from LISTSERVs. These are almost always available only by subscribing to a particular group but have the advantage that many lists have long-term archives.

Finally, institutions such as government agencies and corporations often have large document collections. Corporate collections are usually not available outside the corporation, but government collections often are. One widely studied collection is the MEDLINE data set from the National Institutes of Health, which contains a very large number of abstracts on medical subjects. The advantage of getting documents from such sources is that one can be reasonably sure that the data have been reviewed and are of good quality.

2.2 Document Standardization

Once the documents are collected, it is not uncommon to find them in a variety of different formats, depending on how the documents were generated. For example, some documents may have been generated by a word processor with its own proprietary format; others may have been generated using a simple text editor and saved as ASCII text; and some may have been scanned and stored as images. Clearly, if we are to process all the documents, it's helpful to convert them to a standard format.

The computer industry as a whole, including most of the text-processing community, has adopted XML (Extensible Markup Language) as its standard exchange format, and this is the standard we adopt for our document collections as well. Briefly, XML is a standard way to insert tags onto a text to identify its parts. Although tags can be nested within other tags to arbitrary depth, we will use that capability only sparingly here. We assume that each document is marked off from the other documents in the corpus by having a distinguishing tag at the beginning, such as <DOC>. By XML convention, tags come in beginning and ending pairs. They are enclosed in angle brackets, and the ending tag has a back slash immediately following the opening angle bracket. Within a document, there can be many other tags to mark off sections of the document. Common sections are <DATE>, <SUBJECT>, <TOPIC>, and <TEXT>. We will focus mainly on <SUBJECT>, <TOPIC>, and <TEXT>. The names are arbitrary. They could just as well be <HEADLINE> and <BODY>. An example of an XML document is shown in Fig. 2.1, where the document has a distinguishing tag of <DOC>.

Many currently available corpora are already in this format (e.g., the newer corpora available from Reuters). The main reason for identifying the pieces of a document consistently is to allow selection of those parts that will be used to generate features. We will almost always want to use the part delimited as <TEXT> but may also want to include parts marked <SUBJECT>, <HEADLINE>, or the like. Additionally, for text classification or clustering, one wants to generate features from a

```
<DOC>
<TEXT>
<TITLE>
Solving Regression Problems with Rule-based Classifiers
</TITLE>
<AUTHORS>
<AUTHOR>
Nitin Indurkhya
</AUTHOR>
<AUTHOR>
Sholom M. Weiss
</AUTHOR>
</AUTHORS>
<ABSTRACT>
We describe a lightweight learning method that induces an ensemble
of decision-rule solutions for regression problems. Instead of
direct prediction of a continuous output variable, the method
discretizes the variable by k-means clustering and solves the
resultant classification problem. Predictions on new examples are
made by averaging the mean values of classes with votes that are
close in number to the most likely class. We provide experimental
evidence that this indirect approach can often yield strong
results for many applications, generally outperforming direct
approaches such as regression trees and rivaling bagged regression
trees.
</ABSTRACT>
</TEXT>
</DOC>
```

Fig. 2.1 An XML document

TOPIC section if there is one. Selected document parts may be concatenated into a single string of characters or may be kept separate if one wants to distinguish the features generated from the headline, say, from those generated from the document body, and perhaps weight them differently.

Many word processors these days allow documents to be saved in XML format, and stand-alone filters can be obtained to convert existing documents without having to process each one manually. Documents encoded as images are harder to deal with currently. There are some OCR (optical character recognition) systems that can be useful, but these can introduce errors in the text and must be used with care.

Why should we care about document standardization? The main advantage of standardizing the data is that the mining tools can be applied without having to consider the pedigree of the document. For harvesting information from a document, it is irrelevant what editor was used to create it or what the original format was. The software tools need to read data just in one format, and not in the many different formats they came in originally.

2.3 Tokenization

Assume the document collection is in XML format and we are ready to examine the unstructured text to identify useful features. The first step in handling text is to break the stream of characters into words or, more precisely, *tokens*. This is fundamental to further analysis. Without identifying the tokens, it is difficult to imagine extracting higher-level information from the document. Each token is an instance of a *type*, so the number of tokens is much higher than the number of types. As an example, in the previous sentence there are two tokens spelled "the." These are both instances of a type "the," which occurs twice in the sentence. Properly speaking, one should always refer to the frequency of occurrence of a type, but loose usage also talks about the frequency of a token. Breaking a stream of characters into tokens is trivial for a person familiar with the language structure. A computer program, though, being linguistically challenged, would find the task more complicated. The reason is that certain characters are sometimes token delimiters and sometimes not, depending on the application. The characters space, tab, and newline we assume are always delimiters and are not counted as tokens. They are often collectively called *white space*. The characters ( ) < > ! ? " are always delimiters and may also be tokens. The characters . , : - ' may or may not be delimiters, depending on their environment.

A period, comma, or colon between numbers would not normally be considered a delimiter but rather part of the number. Any other comma or colon is a delimiter and may be a token. A period can be part of an abbreviation (e.g., if it has a capital letter on both sides). It can also be part of an abbreviation when followed by a space (e.g., Dr.). However, some of these are really ends of sentences. The problem of detecting when a period is an end of sentence and when it is not will be discussed later. For the purposes of tokenization, it is probably best to treat any ambiguous period as a word delimiter and also as a token.

The apostrophe also has a number of uses. When preceded and followed by non-delimiters, it should be treated as part of the current token (e.g., isn't or D'angelo). When followed by an unambiguous terminator, it might be a closing internal quote or might indicate a possessive (e.g., Tess'). An apostrophe preceded by a terminator is unambiguously the beginning of an internal quote, so it is possible to distinguish the two cases by keeping track of opening and closing internal quotes.

A dash is a terminator and a token if preceded or followed by another dash. A dash between two numbers might be a subtraction symbol or a separator (e.g., 555-1212 as a telephone number). It is probably best to treat a dash not adjacent to another dash as a terminator and a token, but in some applications it might be better to treat the dash, except in the double dash case, as simply a character.

An example of pseudocode for tokenization is shown in Fig. 2.2. A version of this is available in the accompanying software.

To get the best possible features, one should always customize the tokenizer for the available text—otherwise extra work may be required after the tokens are obtained. The reader should note that the tokenization process is language-dependent. We, of course, focus on documents in English. For other languages, although the general principles will be the same, the details will differ.

2.4 Lemmatization

Once a character stream has been segmented into a sequence of tokens, the next possible step is to convert each of the tokens to a standard form, a process usually

```
Initialize:
   Set Stream to the input text string
   Set currentPosition to 0 and internalQuoteFlag to false
   Set delimiterSet to ' , . ; : ! ? ( ) <>+"\n\t space
   Set whiteSpace to \t\n space
Procedure getNextToken:
   L1: cursor := currentPosition; ch := charAt(cursor);
      If ch = endOfStream then return null; endif
   L2: while ch is not endOfStream nor instanceOf(delimiterSet) do
         increment cursor by 1; ch := charAt(cursor);
      endwhile
      If ch = endOfStream then
         If cursor = currentPosition then return null; endif
      endif
      If ch is whiteSpace then
         If currentPosition = cursor then
            increment currentPosition by 1 and goto L1;
         else
            Token := substring(Stream,currentPosition,cursor-1);
            currentPosition := cursor+1; return Token;
         endif
      endif
      If ch = ' then
         If charAt(cursor-1) = instanceOf(delimiterSet) then
            internalQuoteFlag := true; increment currentPosition by 1; goto L1;
         endif
         If charAt(cursor+1) != instanceof(delimiterSet) then
            increment cursor by 1; ch := charAt(cursor); goto L2;
         elseif internalQuoteFlag = true then
            Token := substring(Stream,currentPosition,cursor-1);
            internalQuoteFlag := false;
         else
            Token := substring(Stream,currentPosition,cursor);
         endif
         currentPosition := cursor+1; return Token;
      endif
      If cursor = currentPosition then
         Token := ch; currentPosition := cursor+1;
      else
         Token := substring(Stream,currentPosition,cursor-1);
         currentPosition := cursor;
      endif
      return Token;
endprocedure
```

Fig. 2.2 Tokenization algorithm

referred to as *stemming* or *lemmatization*. Whether or not this step is necessary is application-dependent. For the purpose of document classification, stemming can provide a small positive benefit in some cases. Notice that one effect of stemming is to reduce the number of distinct types in a text corpus and to increase the frequency of occurrence of some individual types. For example, in the previous sentence, the two instances of "types" would be reduced to the stem "type" and would be counted as instances of that type, along with instances of the tokens "type" and "typed." For classification algorithms that take frequency into account, this can sometimes make a difference. In other scenarios, the extra processing may not provide any significant gains.

2.4.1 Inflectional Stemming

In English, as in many other languages, words occur in text in more than one form. Any native English speaker will agree that the nouns "book" and "books" are two forms of the same word. Often, but not always, it is advantageous to eliminate this kind of variation before further processing (i.e., to normalize both words to the single form "book"). When the normalization is confined to regularizing grammatical variants such as singular/plural and present/past, the process is called "inflectional stemming." In linguistic terminology, this is called "morphological analysis." In some languages, for example Spanish, morphological analysis is comparatively simple. For a language such as English, with many irregular word forms and nonintuitive spelling, it is more difficult. There is no simple rule, for example, to bring together "seek" and "sought." Similarly, the stem for "rebelled" is "rebel," but the stem for "belled" is "bell." In other languages, inflections can take the form of infixing, as, in the German "angeben" (declare), for which the past participle is "angegeben."

Returning to English, an algorithm for inflectional stemming must be part rule-based and part dictionary-based. Any stemming algorithm for English that operates only on tokens, without more grammatical information such as part-of-speech, will make some mistakes because of ambiguity. For example, is "bored" the adjective as in "he is bored" or is it the past tense of the verb "bore"? Furthermore, is the verb "bore" an instance of the verb "bore a hole," or is it the past tense of the verb "bear"? In the absence of some often complicated disambiguation process, a stemming algorithm should probably pick the most frequent choice. Pseudocode for a somewhat simplified inflectional stemmer for English is given in Fig. 2.3. Notice how the algorithm consists of rules that are applied in sequence until one of them is satisfied. Also notice the frequent referrals to a dictionary, usually referred to as a *stemming dictionary*. Although the inflectional stemmer is not expected to be perfect, it will correctly identify quite a significant number of stems. An inflectional stemmer is available with the accompanying software.

2.4.2 Stemming to a Root

Some practitioners have felt that normalization more aggressive than inflectional stemming is advantageous for at least some text-processing applications. The intent of these stemmers is to reach a root form with no inflectional or derivational prefixes

Input: a text token and a dictionary
Doubling consonants: b d g k m n p r l t

Rules:
   If token length < 4 **return** token
   If token is number **return** token
   If token is acronym **return** token
   If token in dictionary **return** the stored stem
   If token ends in s'
      strip the ' and **return** stripped token
   If token ends in 's
      strip the 's and **return** stripped token
   If token ends in "is", "us", or "ss" **return** token
   If token ends in s
      strip s, check in dictionary, and **return** stripped token if there
   If token ends with es
      strip es, check in dictionary, and **return** stripped token if there
   If token ends in ies
      replace ies by y and **return** changed token
   If token ends in s
      strip s and **return** stripped token
   If token doesn't end with ed or ing **return** token
   If token ends with ed
      strip ed, check in dictionary and **return** stripped token if there
   If token ends in ied
      replace ied by y and **return** changed token
   If token ends in eed
      remove d and **return** stripped token if in dictionary
   If token ends with ing
      strip ing (if length > 5) and **return** stripped token if in dictionary
   If token ends with ing and length $\leq$ 5 **return** token
   // Now we have SS, the stripped stem, without ed or ing and it's
   // not in the dictionary (otherwise algorithm would terminate)
   If SS ends in doubling consonant
      strip final consonant and **return** the changed SS if in dictionary
   If doubling consonant was l **return** original SS
   If no doubled consonants in SS
      add e and **return** changed SS if in dictionary
   If SS ends in c or z, or there is a g or l before the final doubling consonant
      add e and **return** changed SS
   If SS ends in any consonant that is preceded by a single vowel
      add e and **return** changed SS
   return SS

Fig. 2.3 Inflectional stemming algorithm

and suffixes. For example, "denormalization" is reduced to the stem "norm." The end result of such aggressive stemming is to reduce the number of types in a text collection very drastically, thereby making distributional statistics more reliable. Additionally, words with the same core meaning are coalesced, so that a concept such as "apply" has only one stem, although the text may have "reapplied", "applications", etc. We cannot make any broad recommendations as to when or when not to use such stemmers. The usefulness of stemming is very much application-dependent. When in doubt, it doesn't hurt to try both with and without stemming if one has the resources to do so.

2.5 Vector Generation for Prediction

Consider the problem of categorizing documents. The characteristic features of documents are the tokens or words they contain. So without any deep analysis of the linguistic content of the documents, we can choose to describe each document by features that represent the most frequent tokens. Figure 2.4 describes this process. A version of this process is available in the accompanying software.

The collective set of features is typically called a *dictionary*. The tokens or words in the dictionary form the basis for creating a spreadsheet of numeric data corresponding to the document collection. Each row is a document, and each column represents a feature. Thus, a cell in the spreadsheet is a measurement of a feature (corresponding to the column) for a document (corresponding to a row). We will soon introduce the predictive methods that learn from such data. But let us first explore the various nuances of this data model and how it might influence the learning methods. In the most basic model of such data, we simply check for the presence or absence of words, and the cell entries are binary entries corresponding to a document and a word. The dictionary of words covers all the possibilities and corresponds to the number of columns in the spreadsheet. The cells will all have ones or zeros, depending on whether the words were encountered in the document.

If a learning method can deal with the high dimensions of such a global dictionary, this simple model of data can be very effective. Checking for words is simple

```
Input:
    ts, all the tokens in the document collection
    k, the number of features desired
Output:
    fs, a set of k features
Initialize:
    hs := empty hashtable

for each tok in ts do
    If hs contains tok then
        i := value of tok in hs
        increment i by 1
    else
        i := 1
    endif
    store i as value of tok in hs
endfor
sk := keys in hs sorted by decreasing value
fs := top k keys in sk
output fs
```

Fig. 2.4 Generating features from tokens

Table 2.1 Dictionary
reduction techniques

Local dictionary
Stopwords
Frequent words
Feature selection
Token reduction: stemming, synonyms

because we do not actually check each word in the dictionary. We build a hash table of the dictionary words and see whether the document's words are in the hash table. Large samples of digital documents are readily available. This gives us confidence that many variations and combinations of words will show up in the sample. This expectation argues for spending less computational time preparing the data to look for similar words or remove weak words. Let the speedy computer find its own way during the learning process.

But, in many circumstances, we may want to work with a smaller dictionary. The sample may be relatively small, or a large dictionary may be unwieldy. In such cases, we might try to reduce the size of the dictionary by various transformations of a dictionary and its constituent words. Depending on the learning method, many of these transformations can improve predictive performance. Table 2.1 lists some of the transformations that can be performed.

If prediction is our goal, we need one more column for the correct answer (or class) for each document. In preparing data for a learning method, this information will be available from the document labels. Our labels are generally binary, and the smaller class is almost always the one of interest. Instead of generating a global dictionary for both classes, we may consider only words found in the class that we are trying to predict. If this class is far smaller than the negative class, which is typical, such a *local dictionary* will be far smaller than the global dictionary.

Another obvious reduction in dictionary size is to compile a list of *stopwords* and remove them from the dictionary. These are words that almost never have any predictive capability, such as articles *a* and *the* and pronouns such as *it* and *they*. These common words can be discarded before the feature generation process, but it's more effective to generate the features first, apply all the other transformations, and at the very last stage reject the ones that correspond to stopwords.

Frequency information on the word counts can be quite useful in reducing dictionary size and can sometimes improve predictive performance for some methods. The most frequent words are often stopwords and can be deleted. The remaining most frequently used words are often the important words that should remain in a local dictionary. The very rare words are often typos and can also be dismissed. For some learning methods, a local dictionary of the most frequent words, perhaps less than 200, can be surprisingly effective.

An alternative approach to local dictionary generation is to generate a global dictionary from all documents in the collection. Special feature selection routines will attempt to select a subset of words that appear to have the greatest potential for prediction. These selection methods are often complicated and independent of the prediction method. Generally, we do not use them and rely on just frequency information, which is quite easy to determine. Any of the feature selection methods that have been used in alternative statistical or machine-learning settings may be tried. Many of these have been developed for real variables and without an emphasis on discrete or binary attributes. Some text-specific methods will be described later on in Sect. 2.5.3, but many of the prediction methods have already been adjusted for text to deal with larger dictionaries rather than repeatedly generating smaller dictionaries. If many classes must be determined, then the generation of a smaller dictionary must be repeated for each prediction problem. For example, if we have 100 topics to categorize, then we have 100 binary prediction problems to solve. Our choices are 100 small dictionaries or one big one. Typically, the vectors implied by a spreadsheet model will also be regenerated to correspond to the small dictionary.

Instead of placing every possible word in the dictionary, we might follow the path of the printed dictionary and avoid storing every variation of the same word. The rationale for this is that all the variants really refer to the same concept. There is no need for singular and plural. Many verbs can be stored in their stem form. Extending the concept, we can also map synonyms to the same token. Of course, this adds a layer of complexity to the processing of text. The gains in predictive performance are relatively modest, but the dictionary size will obviously be reduced. Stemming can occasionally be harmful for some words. If we apply a universal procedure that effectively trims words to their root form, we will encounter occasions where a subtle difference in meaning is missed. The words "exit" and "exiting" may appear to have identical roots, but in the context of programming and error messages, they may have different meanings. Overall, stemming will achieve a large reduction in dictionary size and is modestly beneficial for predictive performance when using a smaller dictionary.

In general, the smaller the dictionary, the more intelligence in its composition is needed to capture the most and best words. The use of tokens and stemming are examples of helpful procedures in composing smaller dictionaries. All these efforts will pay off in improved manageability of learning and perhaps improved accuracy. If nothing else is gained, learning can proceed more rapidly with smaller dictionaries.

Once the set of features has been determined, the document collection can be converted to spreadsheet format. Figure 2.5 shows an example of how this can be done for binary features. An implementation of a similar algorithm is available in the accompanying software. Each column in the spreadsheet corresponds to a feature. For interpretability, we will need to keep the list of features to translate from column number to feature name. And, of course, we will still need the document collection to be able to refer back to the original documents from the rows.

We have presented a model of data for predictive text mining in terms of a spreadsheet that is populated by ones or zeros. These cells represent the presence of the dictionary's words in a document collection. To achieve the best predictive accuracy, we might consider additional transformations from this representation. Table 2.2 lists three different transformations that may improve predictive capabilities.

Word pairs and collocations are simple examples of multiword features discussed in more detail in Sect. 2.5.1. They serve to increase the size of the dictionary but can improve predictive performance in certain scenarios.

Fig. 2.5 Converting
documents to a spreadsheet

```
Input:
  fs, a set of k features
  dc, a collection of n documents
Output: ss, a spreadsheet with n rows and k columns
Initialize: i := 1

for each document d in dc, do
  j := 1
  for each feature f in fs, do
    m := number of occurrences of f in d
    if (m > 0) then ss(row=i, col=j) := 1;
    else ss(row=i, col=j) := 0;
    endif
    increment j by 1
  endfor
  increment i by 1
endfor
output ss
```

Table 2.2 Dictionary feature
transformations

Word pairs, collocations
Frequencies
tf-idf

Instead of zeros or ones as entries in the cells of the spreadsheet, the actual frequency of occurrence of the word could be used. If a word occurs ten times in a document, this count would be entered in the cell. We have all the information of a binary representation, and we have some additional information to contrast with other documents. For some learning methods, the count does give a slightly better result. It also may lead to more compact solutions because it includes the same solution space as the binary data model, yet the additional frequency information may yield a simpler solution. This is especially true of some learning methods whose solutions use only a small subset of the dictionary words. Overall, the frequencies are helpful in prediction but add complexity to the proposed solutions. One compromise that works quite well is to have a three-valued system for cell entries: a one or zero as in the binary representation, with the additional possibility of a 2. Table 2.3 lists the three possibilities, where we map all occurrences of more than two times into a maximum value of 2. Such a scheme seems to capture much of the added value of frequency information without adding much complexity to the model. Another variant involves zeroing values below a certain threshold on the plausible grounds that tokens should have a minimum frequency before being considered of any use. This can reduce the complexity of the spreadsheet significantly and might be a necessity for some data-mining algorithms. Besides simple thresholding, there are a variety of more sophisticated methods to reduce the spreadsheet complexity such as the use of chi-square, mutual information, odds ratio and others. Mutual information can be helpful when considering multiword features.

Table 2.3 Thresholding
frequencies to three values

0—word did not occur
1—word occurred once
2—word occurred 2 or more times

The next step beyond counting the frequency of a word in a document is to modify the count by the perceived importance of that word. The well-known *tf-idf* formulation has been used to compute weightings or scores for words. Once again, the values will be positive numbers so that we capture the presence or absence of the word in a document. In (2.1), we see that the tf-idf weight assigned to word j is the term frequency (i.e., the word count) modified by a scale factor for the importance of the word. The scale factor is called the *inverse document frequency*, which is given in (2.2). It simply checks the number of documents containing word j (i.e., $df(j)$) and reverses the scaling. Thus, when a word appears in many documents, it is considered unimportant and the scale is lowered, perhaps near zero. When the word is relatively unique and appears in few documents, the scale factor zooms upward because it appears important.

$$\text{tf-idf}(j) = \text{tf}(j) * \text{idf}(j), \tag{2.1}$$

$$\text{idf}(j) = \log\left(\frac{N}{df(j)}\right). \tag{2.2}$$

Alternative versions of the basic tf-idf formulation exist, but the general motivation is the same. The net result of this process is a positive score that replaces the simple frequency or binary true-or-false entry in the cell of our spreadsheet. The bigger the score, the more important its expected value to the learning method. Although this transformation is only a slight modification of our original binary-feature model, it does lose the clarity and simplicity of the earlier presentation.

Another variant is to weight the tokens from different parts of the document differently. For example, the words in the subject line of a document could receive additional weight. An effective variant is to generate separate sets of features for the categories (for each category, the set of features is derived only from the tokens of documents of that category) and then pool all the feature sets together.

All of these models of data are modest variations of the basic binary model for the presence or absence of words. Which of the data transformations are best? We will not give a universal answer. Experience has shown that the best prediction accuracy is dependent on mating one of these variations to a specific learning method. The best variation for one method may not be the one for another method. Is it necessary to test all variations with all methods? When we describe the learning methods, we will give guidelines for the individual methods based on general research experience. Moreover, some methods have a natural relationship to one of these representations, and that alone would make them the preferred approach to representing data.

Much effort has been expended in transforming this word model of data into a somewhat more cryptic presentation. The data remain entries in the spreadsheet

Fig. 2.6 Spreadsheet to
sparse vectors

Spreadsheet				Sparse Vectors
0	15	0	3	(2,15) (4,3)
12	0	0	0	(1,12)
8	0	5	2	(1,8) (3,5) (4,2)

cells, but their value may be less intelligible. Some of these transformations are techniques for reducing duplication and dimensions. Others are based on careful empirical experimentation that supports their value in increased predictive capabilities. We will discuss several classes of prediction methods. They tend to work better with different types of data transformations.

Although we describe data as populating a spreadsheet, we expect that most of the cells will be zero. Most documents contain a small subset of the dictionary's words. In the case of text classification, a text corpus might have thousands of word types. Each individual document, however, has only a few hundred unique tokens. So, in the spreadsheet, almost all of the entries for that document will be zero. Rather than store all the zeros, it is better to represent the spreadsheet as a set of *sparse vectors*, where a row is represented by a list of pairs, one element of the pair being a column number and the other element being the corresponding nonzero feature value. By not storing the zeros, savings in memory can be immense. Processing programs can be easily adapted to handle this format. Figure 2.6 gives a simple example of how a spreadsheet is transformed into sparse vectors. All of our proposed data representations are consistent with such a sparse data representation.

2.5.1 Multiword Features

Generally, features are associated with single words (tokens delimited by white space). Although this is reasonable most of the time, there are cases where it helps to consider a group of words as a feature. This happens when a number of words are used to describe a concept that must be made into a feature. The simplest scenario is where the feature space is extended to include pairs of words. Instead of just separate features for *bon* and *vivant*, we could also have a feature for *bon vivant*. But why stop at pairs? Why not consider more general *multiword* features?

The most common example of this is a named entity, for example *Don Smith* or *United States of America*. Unlike word pairs, the words need not necessarily be consecutive. For example, in specifying *Don Smith* as a feature, we may want to ignore the fact that he has a middle name of *Leroy* that may appear in some references to the person. Another example of a multiword feature is an adjective followed by a noun, such as *broken vase*. In this case, to accommodate many references to the noun that involve a number of adjectives with the desired adjective not necessarily adjacent to the noun, we must permit some flexibility in the distance between the adjective and noun. In the same example of the vase, we want to accept a phrase such as *broken and dirty vase* as an instance of *broken vase*. An even more abstract case is when words simply happen to be highly correlated in the text. For instance, in stories about Germany boycotting product Y, the word-stem *German* would be highly correlated with the wordstem *boycott* within a small window (say, five words). Thus, more generally, multiword features consist of x number of words occurring within a maximum window size of y (with $y \geq x$ naturally).

The key question is how such features can be extracted from text. How smart do we have to be in finding such features? Named entities can be extracted using specialized methods. For other multiword features, a more general approach might be to treat them like single-word features. If we use a frequency approach, then we will only include those combinations of words that occur relatively frequently. A straightforward implementation would simply examine all combinations of up to x words within a window of y words. Clearly, the number of potential features grows significantly when multiword features are considered.

Measuring the value of multiword features is typically done by considering correlation between the words in potential multiword features. A variety of measures based on mutual information or the likelihood ratio may be used for this purpose. In the accompanying software, (2.3), which computes an association measure AM for the multiword T, is used for evaluating multiword features, where size(T) is the number of words in phrase T and freq(T) is the number of times phrase T occurs in the document collection.

$$AM(T) = \frac{\text{size}(T) \log_{10}(\text{freq}(T))\text{freq}(T)}{\sum_{word_i \in T} \text{freq}(word_i)}. \tag{2.3}$$

Other variations can occur depending on whether stopwords are excluded before building multiword features.

An algorithm for generating multiword features is shown in Fig. 2.7, which extends Fig. 2.4 to multiword features. A straightforward implementation can consume a lot of memory, but a more efficient implementation uses a sliding window to generate potential multiwords in a single pass over the input text without having to store too many words in memory. A version of this is implemented in the accompanying software.

Generally, multiword features are not found too frequently in a document collection, but when they do occur they are often highly predictive. They are also particularly satisfying for explaining a learning method's proposed solution. The downside to using multiwords is that they add an additional layer of complexity to the processing of text, and some practitioners may feel it's the job of the learning methods to combine the words without a preprocessing step to compose multiword features. However, if the learning method is not capable of doing this, the extra effort may be worthwhile because multiwords are often highly predictive and enhance the interpretability of results.

```
Input:
    ts, sequence of tokens in the document collection
    k, the number of features desired
    mwl, maximum length of multiword
    mws, maximum span of words in multiword
    slvl, correlation threshold for multiword features
    mfreq, frequency threshold for accepting features
Output:
    fs, a set of k features
Initialize:
    hs := empty hashtable

for each tok in ts do
    Generate a list of multiword tokens ending in tok.
    This list includes the single-word tok and uses the inputs mws and mwl.
    Call this list mlist.
    for each mtok in mlist do
        If hs contains mtok then
            i := value of mtok in hs
            increment i by 1
        else
            i := 1
        endif
        store i as value of mtok in hs
    endfor
endfor
sk := keys in hs sorted by decreasing value
delete elements in sk with a frequency < mfreq
delete multiword elements in sk with an association measure < slvl
fs := top k keys in sk
output fs
```

Fig. 2.7 Generating multiword features from tokens

2.5.2 Labels for the Right Answers

For prediction, an extra column must be added to the spreadsheet. This last column of the spreadsheet, containing the label, looks no different from the others. It is a one or zero indicating that the correct answer is either true or false. What is the label? Traditionally, this label has been a topic to index the document. Sports or financial stories are examples of topics. We are not making this semantic distinction. Any answer that can be measured as true or false is acceptable. It could be a topic or category, or it could be an article that appeared prior to a stock price's rise. As long as the answers are labeled correctly relative to the concept, the format is acceptable. Of course, that doesn't mean that the problem can readily be solved. In the sparse vector format, the labels are appended to each vector separately as either a one (positive class) or a zero (negative class).

2.5.3 *Feature Selection by Attribute Ranking*

In addition to the frequency-based approaches mentioned earlier, feature selection can be done in a number of different ways. In general, we want to select a set of features for each category to form a local dictionary for the category. A relatively simple and quite useful method for doing so is by independently ranking feature attributes according to their predictive abilities for the category under consideration. In this approach, we can simply select the top-ranking features.

The predictive ability of an attribute can be measured by a certain quantity that indicates how correlated a feature is with the class label. Assume that we have n documents, and x_j is the presence or absence of attribute j in a document x. We also use y to denote the label of the document; that is, the last column in our spreadsheet model. A commonly used ranking score is the *information gain* criterion, which can be defined as

$$IG(j) = L_{label} - L(j),$$

where

$$L_{label} = \sum_{c=0}^{1} \Pr(y = c) \log_2 \frac{1}{\Pr(y = c)}, \tag{2.4}$$

$$L(j) = \sum_{v=0}^{1} \Pr(x_j = v) \sum_{c=0}^{1} \Pr(y = c | x_j = v) \log_2 \frac{1}{\Pr(y = c | x_j = v)}. \tag{2.5}$$

The quantity $L(j)$ is the number of bits required to encode the label and the attribute j minus the number of bits required to encode the attribute j. That is, $L(j)$ is the number of bits needed to encode the label given that we know the attribute j. Therefore, the information gain $L_{label} - L(j)$ is the number of bits we can save for encoding the class label if we know the feature j. Clearly, it measures how useful a feature j is from the information-theoretical point of view.

Since L_{label} is the same for all j, we can simply compute $L(j)$ for all attributes j and select the ones with the smallest values. Quantities that are needed to compute $L(j)$ in (2.5) can be easily estimated using the following plug-in estimators:

$$\Pr(x_j = v) = \frac{\text{freq}(x_j = v) + 1}{n + 2},$$

$$\Pr(y = c | x_j = v) = \frac{\text{freq}(x_j = v, label = c) + 1}{\text{freq}(x_j = v) + 2}.$$

2.6 Sentence Boundary Determination

If the XML markup for a corpus does not mark sentence boundaries, it is often necessary for these to be marked. At the very least, it is necessary to determine when a period is part of a token and when it is not. For more sophisticated linguistic parsing, the algorithms often require a complete sentence as input. We shall also see other information extraction algorithms that operate on text a sentence at a time. For these algorithms to perform optimally, the sentences must be identified correctly. Sentence boundary determination is essentially the problem of deciding which instances of a period followed by whitespace are sentence delimiters and which are not since we assume that the characters ? and ! are unambiguous sentence boundaries. Since this is a classification problem, one can naturally invoke standard classification software on training data and achieve accuracy of more than 98%. This is discussed at some length in Sect. 2.12. However, if training data are not available, one can use a hand-crafted algorithm.

Figure 2.8 gives an algorithm that will achieve an accuracy of more than 90% on newswire text. Adjustments to the algorithm for other corpora may be necessary to get better performance. Notice how the algorithm is implicitly tailored for English. A different language would have a completely different procedure but would still involve the basic idea of rules that examine the context of potential sentence boundaries. A more thorough implementation of this algorithm is available in the accompanying software.

Input: a text with periods
Output: same text with End-of-Sentence (EOS) periods identified

Overall Strategy:
 1. Replace all identifiable non-EOS periods with another character
 2. Apply rules to all the periods in text and mark EOS periods
 3. Retransform the characters in step 1 to non-EOS periods
 4. Now the text has all EOS periods clearly identified

Rules:
 All ? ! are EOS
 If " or ' appears before period, it is EOS
 If the following character is not white space, it is not EOS
 If) }] before period, it is EOS
 If the token to which the period is attached is capitalized
   and is < 5 characters and the next token begins uppercase,
   it is not EOS
 If the token to which the period is attached has other periods,
   it is not EOS
 If the token to which the period is attached begins with a lowercase
   letter and the next token following whitespace is uppercase,
   it is EOS
 If the token to which the period is attached has < 2 characters,
   it is not EOS
 If the next token following whitespace begins with $ ({ [" ' it is EOS
 Otherwise, the period is not EOS

Fig. 2.8 End-of-sentence detection algorithm

2.7 Part-of-Speech Tagging

Once a text has been broken into tokens and sentences, the next step depends on what is to be done with the text. If no further linguistic analysis is necessary, one might proceed directly to feature generation, in which the features will be obtained from the tokens. However, if the goal is more specific, say recognizing names of people, places, and organizations, it is usually desirable to perform additional linguistic analyses of the text and extract more sophisticated features. Toward this end, the next logical step is to determine the *part of speech* (POS) of each token.

In any natural language, words are organized into grammatical classes or parts of speech. Almost all languages will have at least the categories we would call nouns and verbs. The exact number of categories in a given language is not something intrinsic but depends on how the language is analyzed by an individual linguist.

In English, some analyses may use as few as six or seven categories and others nearly one hundred. Most English grammars would have as a minimum noun, verb, adjective, adverb, preposition, and conjunction. A bigger set of 36 categories is used in the Penn Tree Bank, constructed from the Wall Street Journal corpus discussed later on. Table 2.4 shows some of these categories.

Table 2.4 Some of the categories in the penn tree bank POS set

Tag	Description
CC	Coordinating conjunction
CD	Cardinal number
DT	Determiner
EX	Existential there
FW	Foreign word
IN	Preposition or subordinating conjunction
JJ	Adjective
JJR	Adjective, comparative
JJS	Adjective, superlative
LS	List item marker
MD	Modal
NN	Noun, singular or mass
NNS	Noun, plural
POS	Possessive ending
UH	Interjection
VB	Verb, base form
VBD	Verb, past tense
VBG	Verb, gerund or present participle
VBN	Verb, past participle
VBP	Verb, non-3rd person singular present
WDT	Wh-determiner

Dictionaries showing word—POS correspondence can be useful but are not sufficient. All dictionaries have gaps, but even for words found in the dictionary, several
parts of speech are usually possible. Returning to an earlier example, "bore" could
be a noun, a present tense verb, or a past tense verb. The goal of POS tagging is to
determine which of these possibilities is realized in a particular text instance.

Although it is possible, in principle, to manually construct a part-of-speech tagger, the most successful systems are generated automatically by machine-learning
algorithms from annotated corpora. Almost all POS taggers have been trained on
the Wall Street Journal corpus available from LDC (Linguistic Data Corporation,
www.ldc.upenn.edu) because it is the most easily available large annotated corpus.
Although the WSJ corpus is large and reasonably diverse, it is one particular genre,
and one cannot assume that a tagger based on the WSJ will perform as well on, for
example, e-mail messages. Because much of the impetus for work on information
extraction has been sponsored by the military, whose interest is largely in the processing of voluminous news sources, there has not been much support for generating
large training corpora in other domains.

2.8 Word Sense Disambiguation

English words, besides being ambiguous when isolated from their POS status, are
also very often ambiguous as to their meaning or reference. Returning once again to
the example "bore," one cannot tell without context, even after POS tagging, if the
word is referring to a person—"he is a bore"—or a reference to a hole, as in "the
bore is not large enough." The main function of ordinary dictionaries is to catalog
the various meanings of a word, but they are not organized for use by a computer
program for disambiguation. A large, long-running project that focused on word
meanings and their interrelationships is Wordnet, which aimed to fill in this gap. As
useful as Wordnet is, by itself it does not provide an algorithm for selecting a particular meaning for a word in context. In spite of substantial work over a long period of
time, there are no algorithms that can completely disambiguate a text. In large part,
this is due to the lack of a huge corpus of disambiguated text to serve as a training
corpus for machine-learning algorithms. Available corpora focus on relatively few
words, with the aim of testing the efficacy of particular procedures. Unless a particular text-mining project can be shown to require word sense disambiguation, it is
best to proceed without such a step.

2.9 Phrase Recognition

Once the tokens of a sentence have been assigned POS tags, the next step is to
group individual tokens into units, generally called phrases. This is useful both for
creating a "partial parse" of a sentence and as a step in identifying the "named
entities" occurring in a sentence, a topic we will return to in greater detail later

on. There are standard corpora and test sets for developing and evaluating phrase recognition systems that were developed for various research workshops. Systems are supposed to scan a text and mark the beginnings and ends of phrases, of which the most important are noun phrases, verb phrases, and prepositional phrases. There are a number of conventions for marking, but the most common is to mark a word inside a phrase with I-, a word at the beginning of a phrase adjacent to another phrase with B- and a word outside any phrase with O. The I- and B- tags can be extended with a code for the phrase type: I-NP, B-NP, I-VP, B-VP, etc. Formulated in this way, the phrase identification problem is reduced to a classification problem for the tokens of a sentence, in which the procedure must supply the correct class for each token.

Performance varies widely over phrase type, although overall performance measures on benchmark test sets are quite good. A simple statistical approach to recognizing significant phrases might be to consider multiword tokens. If a particular sequence of words occurs frequently enough in the corpora, it will be identified as a useful token.

2.10 Named Entity Recognition

A specialization of phrase finding, in particular noun phrase finding, is the recognition of particular types of proper noun phrases, specifically persons, organizations, and locations, sometimes along with money, dates, times, and percentages. The importance of these recognizers for intelligence applications is easy to see, but they have more mundane uses, particularly in turning verbose text data into a more compact structural form.

From the point of view of technique, this is very like the phrase recognition problem. One might even want to identify noun phrases as a first step. The same sort of token-encoding pattern can be used (B-person, B-location, I-person, etc.), and the problem is then one of assigning the correct class code to each token in a sentence. We shall discuss this problem in detail in Chap. 6 on information extraction.

2.11 Parsing

The most sophisticated kind of text processing we will consider, briefly, is the step of producing a full parse of a sentence. By this, we mean that each word in a sentence is connected to a single structure, usually a tree but sometimes a directed acyclic graph. From the parse, we can find the relation of each word in a sentence to all the others, and typically also its function in the sentence (e.g. subject, object, etc.). There are very many different kinds of parses, each associated with a linguistic theory of language. This is not the place to discuss these various theories. For our purposes, we can restrict attention to the so-called "context-free" parses. One can envision a parse of this kind as a tree of nodes in which the leaf nodes are the words of a sentence, the phrases into which the words are grouped are internal nodes, and

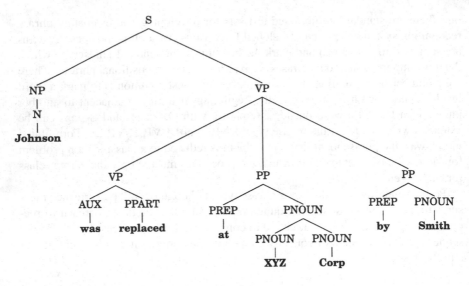

Fig. 2.9 Simple parse tree

```
Johnson was replaced at XYZ Corp. by Smith .
-----------------------------------------------------------------
.----- subj(n)       Johnson1(1)       noun propn sg h m gname sname
o----- top           be(2,1,3)         verb vfin vpast sg vsubj
'----- pred(en)      replace1(3,7,1,u) verb ven vpass
  '--- vprep         at1(4,6)          prep pprefv staticp
  |  '- objprep(n)   XYZ Corp.1(6)     noun propn sg glom ctitle
  '--- subj(agent)   by1(7,8)          prep pprefv
    '- objprep(n)    Smith1(8)         noun propn sg h sname
```

Fig. 2.10 Parse tree—English Slot Grammar

there is one top node at the root of the tree, which usually has the label S. There are a number of algorithms for producing such a tree from the words of a sentence. Considerable research has been done on constructing parsers from a statistical analysis of tree banks of sentences parsed by hand. The best-known and most widely used tree bank is of parsed sentences from the Wall Street Journal and is available from LDC.

The reason for considering such a comparatively expensive process is that it provides information that phrase identification or partial parsing cannot provide. Consider a sentence such as "Johnson was replaced at XYZ Corp. by Smith." for which a simple parse is shown in Fig. 2.9.

From the linear order of phrases in a partial parse, one might conclude that Johnson replaced Smith. A parse that identifies the sentence as passive has information allowing the extraction of the correct "Smith replaced Johnson." An example of a parse giving more information than a simple tree is shown in Fig. 2.10. The parse is the output of the English Slot Grammar. In this example, the tree is drawn, using printer characters, on its side, with the top of the tree to the left. Notice in particular that the "by" phrase containing "Smith" is identified as the agent, although "Johnson" is marked as the subject.

2.12 Feature Generation

Although we emphasized that our focus is on statistical methods, the reason for the linguistic processing described earlier is to identify features that can be useful for text mining. As an example, we will consider how good features depend on the type of object to be classified and how such features can be obtained using the processes discussed so far.

Let us consider the problem of segmenting a text into sentences. A hand-crafted algorithm for this task was given earlier. However, let's say we want to learn a set of similar rules from training data instead. What sorts of features would we generate? Since the object to be classified is a period, each feature vector corresponds to a period occurring in the text. Now we need to consider what characteristics of the surrounding text are useful features. From the algorithm, we can see that the useful features are the characters or character classes near the period, including the character of the token to which the period is attached and the characters of the following token. Therefore, the necessary linguistic processing only involves tokenization of the text.

A more sophisticated example would be the identification of the part of speech (POS) of each word in a text. This is normally done on a text that has first been segmented into sentences. The influence of one word on the part-of-speech of another does not cross sentence boundaries. The object that a feature vector represents is a token. Features that might be useful in identifying the POS are, for example, whether or not the first letter is capitalized (marking a proper noun), if all the characters are digits, periods, or commas (marking a number), if the characters are alternating uppercase letters and periods (an abbreviation), and so on. We might have information from a dictionary as to the possible parts of speech for a token. If we assume the POS assignment goes left to right through a sentence, we have POS assignments of tokens to the left as possible features. In any case, we have the identity of tokens to the left and to the right. For example, "the" most likely precedes either a noun or an adjective. So, for this task, we basically need tokenization plus analysis of the tokens, plus perhaps some dictionaries that tell what the possibilities are for each token in order to create a feature vector for each token.

The feature vector for a document is assigned a particular class (or set of classes). The feature vector for classifying periods as End-Of-Sentence or not is assigned to one of two classes. The feature vector for POS assignment has one of a finite set of classes. The class of the feature vector for each token in the partial parsing task was outlined above. These classes are not intrinsic or commonly agreed properties of a token. They are invented constructs specific to the problem. Let us consider what features are important for the feature vector for partial parsing. Token identity is clearly one of these, as is the token POS. Additionally, the identity and POS of the tokens to the left and right of the token whose vector is being constructed are important features. So is the phrasal class of tokens to the left that have already been assigned. Sentence boundaries are particularly important since phrases do not cross them. Individual token features, on the other hand, are not important because they have already been taken into account for POS assignment.

For named entity detection, the same kind of token class encoding scheme can be used as in the chunking task (i.e., B-Person, I-Person, etc.). All the named entities are noun phrases, so it is possible but not necessary that a sentence will first be segmented into noun phrases. This might result in unnecessary work since named entities are typically made up of proper nouns. For this task, dictionaries can be particularly important. One can identify tokens as instances of titles, such as "Doctor" or "President," providing clues as to the class of proper noun phrases to the right. Other dictionaries can list words that are typically the end of organization names, like "Company," "Inc.," or "Department." There are also widely available gazetteers (i.e., lists of place names and lists of organization names). Identifying a token as being a piece of such a dictionary entry is useful but not definitive because of ambiguity of names (e.g., "Arthur Anderson" might be referring to a person or to a company). Besides the dictionary information, other useful features are POS, sentence boundaries, and the class of tokens already assigned.

2.13 Summary

Documents are composed of words, and machine learning methods process numerical vectors. This chapter discusses how words are transformed into vectors, readying them for processing by predictive methods. Documents may appear in different formats and may be collected from different sources. With minor modifications, they can be organized and unified for prediction by specifying them in a standard descriptive language, XML. The words or tokens may be further reduced to common roots by stemming. These tokens are added to a dictionary. The words in a document can be converted to vectors using local or global dictionaries. The value of each entry in the vector will be based on measures of frequency of occurrence of words in a document such as term frequency (tf and idf). An additional entry in a document vector is a label of the correct answer, such as its topic. Dictionaries can be extended to multiword features like phrases. Dictionary size may be significantly reduced by attribute ranking. The general approach is purely empirical, preparing data for statistical prediction. Linguistic concepts are also discussed including part-of-speech tagging, word sense disambiguation, phrase recognition, parsing and feature generation.

2.14 Historical and Bibliographical Remarks

A detailed account of linguistic processing issues can be found in Indurkhya and Damerau (2010) and Jurafsky and Martin (2008). Current URLs for organizations such as LDC, ICAME, TEI, the Oxford Text Archive, and the like are easily found through a search engine. The new Reuters corpus, RCV1, is discussed in Lewis *et al.* (2004) and is available directly from Reuters. The older Reuters-21578 Distribution 1.0 corpus is also available on the Internet at several sites. Using a search engine with the query "download Reuters 21578" will provide a list of a number of sites where this corpus can be obtained. There are a number of Web sites that have many links to corpora in many languages. Again, use of a search engine with the query "corpus linguistics" will give the URLs of active sites. There are many books on XML; for example Ray (2001). The best-known algorithm for derivational regularization is the Porter stemmer (Porter 1980), which is in the public domain. At the time of this writing, it can be downloaded from http://www.tartarus.org/~martin/PorterStemmer. ANSI C, Java, and Perl versions are available. Another variation on stemming is to base the unification of tokens into stems on corpus statistics (i.e., the stemming is corpus based) (Xu and Croft 1998). For information retrieval, this algorithm is said to provide better results than the more aggressive Porter stemmer.

For end-of-sentence determination, Walker *et al.* (2001) compares a hard-coded program, a rule-based system, and a machine-learning solution on the periods and some other characters in a collection of documents from the Web. The machine-learning system was best of the three. The F-measures (see Sect. 3.5.1) were 98.37 for the machine-learning system, 95.82 for the rule-based system, and 92.45 for the program. Adjusting a machine-learning solution to a new corpus is discussed in Zhang *et al.* (2003).

Examples of part-of-speech taggers are Ratnaparkhi (1995) and Brill (1995). The Brill tagger is in the public domain and is in wide use.

For a survey of work on word sense disambiguation, see Ide and Véronis (1998). Wordnet is discussed in Feldbaum (1998). The database and a program to use it can be obtained from the Internet at http://wordnet.princeton.edu.

Phrase recognition is also known as "text chunking" (Sang and Buchholz 2000). A number of researchers have investigated this problem as classification, beginning with Ramshaw and Marcus (1995). A variety of machine-learning algorithms have been used: support vector machines (Kudoh and Matsumoto 2000), transformation-based learning (Ramshaw and Marcus 1995), a linear classifier (Zhang *et al.* 2002), and others. Many more details can be found at http://ifarm.nl/signll/conll.

Work on named entity recognition has been heavily funded by the US government, beginning with the Message Understanding Conferences and continuing with the ACE project Further details can be obtained from the following sites:

http://www.itl.nist.gov/iaui/894.02/related_projects/muc/index.html
http://www.itl.nist.gov/iaui/894.01/tests/ace/index.html.

A number of named entity recognition systems are available for license, such as the Nymble system from BBN (Bikel *et al.* 1997). State of the art is about 90% in recognition accuracy.

Algorithms for constructing context-free parse trees are discussed in Earley (1970) and Tomita (1985). Constructing parsers from tree-banks is discussed in

Charniak (1997), Ratnaparkhi (1999), Chiang (2000), and others. A description of English Slot Grammar can be found in McCord (1989). A number of organizations have posted demo versions of their parsers on Web sites. These can be used to compare the output of different parsers on the same input sentence. Three examples of such sites are:

http://www.agfl.cs.ru.nl/EP4IR/english.html
http://www.link.cs.cmu.edu/link/submit-sentence-4.html
http://www2.lingsoft.fi/cgi-bin/engcg.

Early work in feature generation for document classification includes (Lewis 1992; Apté *et al.* 1994), and others. Tan *et al.* (2002) showed that bigrams plus single words improved categorization performance in a collection of Web pages from the Yahoo Science groups. The better performance was attributed to an increase in recall. A special issue of the *Journal of Machine Learning Research*, in 2003 was devoted to feature selection and is available online. One of the papers (Forman 2003) presents experiments on various methods for feature reduction. A useful reference on word selection methods for dimensionality reduction is Yang and Pedersen (1997), which discusses a wide variety of methods for selecting words useful in categorization. It concludes that document frequency is comparable in performance to expensive methods such as information gain or chi-square.

2.15 Questions and Exercises

1. How would you evaluate a tokenizer?
2. What component in the software set would need to be replaced in order to run a classifier for periods as end-of-sentence?
3. Create a properties file for mkdict for Reuters-21578 train data set with no special parameter settings and run using a size of 500. Don't forget to set all the tags (especially bodytags and doctag) in the properties file and ensure that it is in the current directory where the java command is executed.
4. Modify the parameters in the properties file to use stop words and stemming. You can use the files provided in ExerciseFiles.zip that you downloaded earlier. Run and note the differences in the dictionary from the previous exercise.
5. Generate a local dictionary for the category *earn*. This dictionary will be built from documents that have the topic *earn*. Check the dictionary file. Lets say you don't want any of the numeric features or ones that include the characters #, & and -. You could edit the file and delete these manually. Or else you could edit the properties file and regenerate the local dictionary. (Hint: add the characters you want to exclude to BOTH whitespace-chars and word-delimiters.) Note that the dictionary is still not perfect, but we will let the prediction programs decide which features to use.
6. Use the dictionary from the previous exercise to generate vectors for the category *earn* for both the training and test data sets.

Chapter 3
Using Text for Prediction

The words *prediction* and *forecast* conjure up images of momentous decisions and complex processes fraught with inaccuracies. From a statistical perspective, it's a straightforward problem that has a solution. Of course, the solution may not always be very good. The problem presents itself as in Fig. 3.1. Given a sample of examples of past experience, we project to new examples. If the future is similar to the past, we may have an opportunity to make accurate predictions. An example of such a situation is where one tries to predict the future share price of a company based on historical records of the company's share price and other measures of its performance.

Making a prediction requires more than a lookup of past experience. Even if we effectively characterize these experiences in a consistent way, the test of success is on new examples. For prediction, a pattern must be found in past experience that will hold in the future, leading to accurate results on new, unseen examples. If a new example presents itself in a form that is radically different from prior experience, learning from past experience will prove inadequate. Machine learning and statistical methods do not learn from basic principles. They have no ability to reason and reach new conclusions for new situations. Still, perhaps surprisingly, many prediction problems can be solved by finding patterns in prior experience. If samples can be obtained and organized in the right format, finding patterns is almost effortless, even in very high-dimensional feature spaces.

The classical prediction problem for text is called *text categorization*. Here, a set of categories is predefined, and the objective is to assign a category or topic to a new document. For example, we can collect newswire articles and describe a set of topics such as financial or sports stories. The set of topics is fixed. When news arrives, the words are examined, and articles are assigned topics from a fixed list of possible topics.

However, characterizing all prediction from text as text categorization is too narrow. Prediction from text can be just as ambitious as prediction for numerical data mining. In statistical terms, prediction has a very specific characterization, and it need not deal with just topic assignment to documents. Prediction for text follows the classical lines of all numerical classification problems, and we can use a tradi-

S.M. Weiss et al., *Fundamentals of Predictive Text Mining,*
Texts in Computer Science 41,
DOI 10.1007/978-1-84996-226-1_3, © Springer-Verlag London Limited 2010

Fig. 3.1 Predicting the future
based on the past

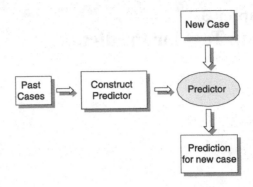

...	unsubscribe	...	enlargement	...	ink	...	spam
...	yes	...	yes	...	yes	...	true
...	no	...	no	...	no	...	false
...	...	...	...	...	...	...	...

Fig. 3.2 Abstract spreadsheet for spam prediction

...	profits	...	increased	...	earnings	...	stock-price
...	yes	...	yes	...	yes	...	1
...	yes	...	no	...	yes	...	0
...	...	...	...	...	...	...	...

Fig. 3.3 Abstract spreadsheet for predicting stock price

tional model of data that applies to any sampling application where the answers are presented as true or false.

The prediction problem for text is generally a classification problem. In our spreadsheet model of data, we have the usual rows and columns, where a row is an example and a column is an attribute to be measured. For classification, we have an additional column, a label identifying the correct answer. For text, the answer is something that is true or false. For example, the labels for a stock price prediction problem may be a 1 for a stock price that goes up and a 0 for unchanged or down.

Figures 3.2 and 3.3 are abstract templates of spreadsheets that might be composed for prediction. Figure 3.2 is the classical text categorization application, where the goal is to filter spam e-mail from valid e-mail. Figure 3.3 illustrates that document classification might be explicitly predictive. The examples are words found in news stories about companies, and the labels are whether the stock price rose in some time period following the article.

So far, we have not shied away from describing text as unstructured data that can be converted into structured data, where classical machine-learning methods can be applied. There remain many nuances in the recipe that do not alter this worldview but can make our trip to obtaining good results more direct. Let's look at predictive methods from the perspective of text and our experience in choosing the best route for their application.

3.1 Recognizing that Documents Fit a Pattern

What kinds of documents are we talking about? These are generally documents in digital form that can be accessed by computer. While they could be as large as a book or a manual, they are more likely to be relatively brief. The prototypical digital document is a newswire or magazine article. Larger documents, such as books, are more structured and are composed of smaller subcomponents such as chapters.

If we pool all the words in a document collection, we often have a large dictionary representing the complete set of words appearing in all documents. It is not unusual for the global dictionary to have hundreds of thousands of words. Yet a local dictionary, taken relative to documents only for a single topic, can be far smaller. As seen in the previous chapter, given a document and a dictionary, we can encode the document as a vector of numbers, in the simplest form a vector of ones and zeros, representing the presence or absence of individual words. Most of the entries will be zero. This concept of sparseness of words for a document or topic is consistent with the notion of finding a pattern of words that is characteristic of a label. To be successful in prediction, we expect to find common characteristics—patterns of words in documents. Once again, these patterns will refer to only a small space of the words that occur in the subset of documents with the same label.

The spreadsheet of Fig. 3.4 is an idealized view of words that form a pattern. For prediction, we examine patterns relative to a label, where the label is the correct answer. The label is true or false, represented by a zero or one in the last column. We see that the same two words (Word2 and Word4) always occur for class 1 and never occur together for class 0. A pattern is formed when a combination of words occurs for the class of interest and not for the negative class.

For all applications of data mining, the accuracy of predictions is dependent on the predictive quality of the attributes. For text mining, these are words or stemmed words. Not all labels can be discriminated. Even people can have a tough time predicting where a stock price may move in a week, whereas they may find it easier to

word1	word2	word3	word4	word5	...	wordN	label
0	1	1	1	0	...	1	1
1	0	1	0	0	...	1	0
1	1	0	1	1	...	0	1
0	1	0	0	1	...	1	0
1	1	1	1	0	...	1	1
0	1	0	1	1	...	0	1
1	0	1	1	0	...	1	0
0	1	1	1	1	...	0	1

Fig. 3.4 Predictive patterns in a spreadsheet

word1	word2	word3	word4	word5	...	wordN	label
0	1	0	1	0	...	1	1
1	0	1	0	0	...	0	0
1	0	0	1	1	...	0	1
0	1	0	0	1	...	1	0
1	0	0	1	0	...	1	1
0	1	0	0	1	...	0	1
0	0	1	1	0	...	0	0
0	0	1	0	1	...	0	1

Fig. 3.5 Spreadsheet with no obvious patterns

forecast their income for the next week. In contrast to Fig. 3.4, in Fig. 3.5, we see no obvious predictors for class 1. No simple pattern is found that separates the two classes. For an ideal pattern using only a single word, a 1 is found in the word's cell when the class is one and a 0 when the class is zero.

3.2 How Many Documents Are Enough?

Predictive text mining needs samples of prior experience. From these samples, a method learns how to make predictions on new documents. The classic application is text categorization where newswire stories are classified by topic. Because any story may be assigned more than one topic, the application is decomposed into many binary problems, one for each topic. Although it is possible to assemble a collection of many thousands of documents, the sample of documents for building predictors must come with labels. This initial assignment of labels is likely to require human intervention, and the effort may be a time-consuming process. We may think of text mining as a completely automated process, but, in reality, the assignment of labels is a bottleneck.

The collection of documents may be very large, yet the number of examples for some topics may be very small. If it's a rare topic, it may be so distinctive that only a few documents will be sufficient to capture its distinctions from other topics. If the topic is not obviously different from others, then additional documents for that topic may be needed, an objective that is sometimes difficult to meet.

Even after assembling a representative sample covering all topics, the task of acquiring data is not complete. A document collection may evolve over time. News stories have a short life span. The popular stories of one year may be of no interest in later years. The topics may be relatively stable, but the representative documents for a topic may vary over time. Even if the sample is changing only slowly, provisions must be made to relearn from the data.

In contrast to many other applications of data mining, text mining provides advantages to the developers of predictive models. For many data-mining applications, the developers process the data but have only a superficial understanding of the measurements. They accept what they are given by the domain experts and do not have a deep understanding of the measurements or their relationship with each other. Results are analyzed primarily by empirical analysis. When something goes awry, we may have difficulty in attributing this to problems with the collection process or the specification of the features. For text mining, we are much closer to understanding the data, and we all have some expertise. The document is text. We can read and comprehend it, and we analyze a result by going directly to the documents of interest. We can also look at the words forming a pattern for prediction and make judgments about the relevance of those words for a specific prediction class.

How many cases are enough to learn from the sample? Given the characteristics of our sample as outlined above, we cannot directly answer that question. Instead, we depend on correct evaluation of proposed solutions, giving us a good idea of their future performance. We are particularly concerned in the early stages of learning from a small sample, anticipating a need for more documents. Some learning methods may be augmented by artificial cases or human-engineered decision rules. Our main focus is on learning and feedback from an evaluation of results without anticipating the likely success of learning from any size sample.

3.3 Document Classification

Classification is a well-understood problem. A sample is collected. The data are organized in a structured format. Every example is measured in the same way. The answer is expressed in terms of true or false, a binary decision. In mathematical terms, a solution is a function that maps examples to labels, $f : w \rightarrow L$, where w is a vector of attributes and L is a label. In our case, the attributes are words or tokens. The labels can be a goal that is potentially related to the words. Most prior research has been done on indexing, where the label is a broad topic, such as categorizing a document as a sports story. But the label could be anything from a security threat to the direction of stock price movements.

Figure 3.6 illustrates the task in terms of spreadsheets. The first spreadsheet is complete and all cells are filled in. The second spreadsheet has the identical format. The columns have the same meaning in both spreadsheets, the examples are different. Missing from the second spreadsheet are the values of the last column. The

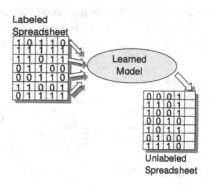

Fig. 3.6 Learning and applying models

objective for a learning method is to learn from the first spreadsheet some model that can predict the values of the last column for the second spreadsheet. Because the examples or rows are not identical, the method cannot employ direct lookup of the first spreadsheet. It needs to find a generalization of the first spreadsheet that will hold on the new examples in the second spreadsheet.

In Fig. 3.6, we learn from the first spreadsheet and apply the inferred model to a new spreadsheet. In statistical terms, we draw a sample from a population, learn from it, and then apply the induced model to new, unlabeled examples drawn from the same population. In classification theory, a new sample and its examples are expected to be *i.i.d.*, independently and identically drawn. In many situations of text mining, this assumption cannot be expected to hold. If we sample newswires, we know that the types of stories will change over time. The population is not stationary; it changes. Still, we can apply classification methods, and with a proper data representation and appropriate evaluation techniques, these learning methods can be effective. We expect that the nature of the documents will remain relatively stable over some time frame, sometimes for a relatively short window of time. For text categorization, the topics and the rules for their assignment will typically not vary greatly over weeks or even months, even though the "hot" topics in the news might change. Over longer periods, documents in the training sample may be discarded and new ones added, and the learning process may then be repeated.

3.4 Learning to Predict from Text

We start with a sample of documents. The initial collection of documents has been mapped into a spreadsheet. From the spreadsheet representation, we expect to learn some decision criteria to classify new examples. Figure 3.7 describes the overall process.

The measurements that we consider have their roots among the well-known classification methods for learning from data. Are we going to review the massive literature of classification-learning methods? Not at all. The many experiments reported in the research literature point us in the direction of the best methods for text. Moreover, these methods have been developed to work with sparse data where the spreadsheet cells are mostly zeros.

We are using editorial license to choose the most interesting and effective predictive methods. In addition, we are choosing specific styles and formats for preparing the data. Thus, we may run one method with a binary representation for words and

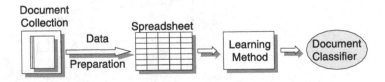

Fig. 3.7 From text to classifiers

another method with tf-idf transformations. Why do we choose a specific combination of technique and representation? We and other researchers have some experience with this issue. You may try every variation, but we have reached a stage where we consistently apply a method in a specialized way.

The four types of methods can be differentiated by the distinctive representations of their solutions. These are the nearest-neighbor methods, logic methods, probabilistic methods, and weighted-scoring methods. We will present a learning method from each of these categories not because we are democratic but simply because these are the most widely used and effective methods. Our problem may be very high-dimensional, but it is also more specialized. The methods will be adapted and in most cases simplified to take advantage of our expectations for text. Let us examine these methods in more detail.

3.4.1 Similarity and Nearest-Neighbor Methods

Finding the nearest neighbors of a document is a process with which we are all familiar. When using a search engine, we present some key words, and the search engine returns its nearest-matching documents, This is a specific type of information retrieval where the new document is only a few key words. The more general problem is to take a new unlabeled document and predict its label. Why not just look up the words of the new document in a stored repository of labeled documents? If the new document consists of a few words, it is likely that some stored document will match those words. That's what happens with a search engine where many documents are returned to match the key words.

Our problem is a variation on this same theme of matching a new document to old ones. In the context of text mining, the problem is one of information retrieval; the more general approach is called nearest-neighbor methods. A complete document will have many words, and it is unlikely that it will completely match a stored document. Instead of an exact match, we try to find the closest matches to the stored documents. We might pull out, for example, the ten best matches, look at their labels, and pick the label that occurs most frequently. This simple process, shown in Fig. 3.8, is the basic algorithm for nearest-neighbor methods. In fact, it is one of the most prominent approaches to prediction for text.

In our example, we decided to pull out the ten best matches. This gets generalized to pulling out the k best matches. But how do we determine k? At one extreme, k could be 1 and we could always take the single most similar document. In general though, looking at a group of similar documents will give a more accurate answer.

Fig. 3.8 Basic nearest-neighbor algorithm for documents

1. Compute the similarity of newDoc to all documents in collection {D(I)}.
2. Select the k documents that are most similar to newDoc.
3. The answer is the label that occurs most frequently in the k selected documents.

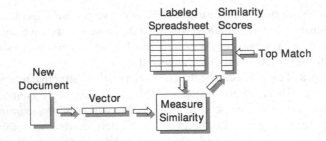

Fig. 3.9 Finding similar documents

Do you usually rely solely on a search engine's top answer? Often, the best answer occurs in the top ten or so matches. The formal name for the method is *k*-nearest neighbors, and finding *k* makes life somewhat more difficult. *k* can be estimated by experimental procedures. These will be discussed in Sect. 3.5, where we review evaluation of performance. In most situations, we will have many binary prediction problems to solve. In text categorization, we have many topics to assign. To determine a distinct *k* for each topic would be a tedious task indeed. In practice, a single value of *k* would be used for almost all categories, with a smaller *k* for those topics where the number of positive examples is less than *k*.

How hard is it to find documents that are similar to a new document? Our documents have been transformed to spreadsheet data. Each document is now a vector of numbers. Figure 3.9 is a graphic of the overall process. The new document is embodied in a vector. That vector is compared to all the other vectors, and a score for similarity is computed.

3.4.2 Document Similarity

Most nearest-neighbor applications compare two examples by measuring the distance between two examples. Equation (3.1) is a general distance measure used to compare two examples. This is simply the difference between each attribute squared. Absolute values could also be used. The larger the distance, the weaker the connection between the examples.

$$\text{distance}(x, y) = (x_1 - y_1)^2 + \cdots + (x_m - y_m)^2. \tag{3.1}$$

For text, a more natural measure is used, which is called *similarity*. When are two documents similar? One view is that they share the same words. The most elementary measure of similarity is to count the number of words that two documents have in common. We can then rank the document by similarity, where the most similar documents have the most shared words. Notice that unlike traditional metrics for computing distance, this one ignores words that occur in one document but not in the other. We are only interested in words that occurred in *both* documents. Traditional nearest-neighbor methods usually do not do well with very high dimensions

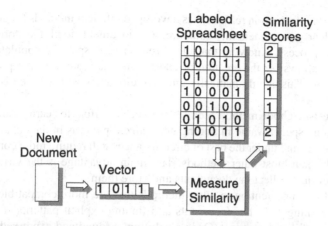

Fig. 3.10 Computing similarity scores for documents

of attributes, in our case thousands of words, where most of the spreadsheets are populated with zeros. One reason for this is the inappropriate distance measure they tend to use. By using similarity as a metric, nearest-neighbor methods become quite practical and effective.

Using a spreadsheet, Fig. 3.10 is an example of a computation of similarity. When a new document is presented, it is compared to every document in the stored collection. Once in spreadsheet format, the comparison may seem time-consuming, but actually it is fast and easy. For every positive word in the new document, we count their number of occurrences in the stored documents. Similarity, S, is the number of positive words found in both a stored document and the new document. With thousands of dictionary words, the number of attributes is huge, creating a very difficult task for these methods. The use of only positive words for similarity seems to eliminate the noise of the zeros, giving extra justification for using the sparse representation.

Is this the best similarity measure for document comparisons? Earlier, we described several ways of representing the measurements in the spreadsheets, including binary, frequency, and tf-idf. All are variations on the same theme, where the additional complexity is an attempt to weigh the quality and importance of the measure. In practice, using the tf-idf variation will usually give the best predictive results for similarity methods. The actual computation of distance in this case is called the *cosine similarity* and has been widely used for information retrieval. We will wait until Chap. 4 to amplify on cosine similarity. This is not the only measure that has been used with good results for many applications. The weighting of tf-idf with a normalization constant involving document length seems to help in achieving somewhat better results than a pure count of shared words.

Information retrieval traditionally deals with finding relevant documents in response to a query. A search engine is the embodiment of modern information retrieval. Yet, the same ideas can be used for another purpose. Here the specific documents retrieved are not our prime interest, but their labels are of paramount importance.

Research on information retrieval has developed efficient methods for computing similarity. Clearly, to sequentially compare new documents to all the stored ones is an inefficient process. Knowing that the spreadsheet is sparsely populated, being mostly zeros, allows for the creation of additional data structures that point to the positive entries. This can dramatically improve efficiency and will be discussed in Chap. 4.

The nearest-neighbor method requires no special effort to learn from the data and provides no special value in finding generalized patterns in the data. It's just a retrieval program and under the best of circumstances will require more computation time to apply than most other methods. The main advantage is the virtually zero training effort: just collect the documents and store them.

We now turn our attention to other methods that are more compatible with the concept of learning from the documents and finding explicit patterns. Their solutions are insightful and compact. Once a solution is found and expressed as a new model, the original sample has a diminished role and is possibly discarded. The most intuitively appealing types of solutions are based on decision rules. We look at these next.

3.4.3 Decision Rules

Just as the nearest-neighbor method can be viewed as a special invocation of a search engine, another method, decision rules, can also be viewed from the same perspective. For the nearest-neighbor method, the search string is the new, unlabeled document, and the most similar documents are retrieved. Only after retrieval are the labels examined to assign the class label of the new document. Although of paramount importance to classification, the labels are secondary to the search process.

How about reversing the search process? Look at the documents that are retrieved, and then generate the search string that would retrieve exactly these documents. That's precisely what a decision rule method attempts to do. The initial sample is a collection of labeled documents. The positive documents are examples of documents that would be retrieved from some hypothetical search string (or strings), and the negative examples are documents that should *not* be retrieved. The problem is to find one or more patterns that would produce those positive examples. The patterns are the same search strings that we employ when we use a search engine: a phrase of one or more words that must occur together to match a document. These patterns are therefore the *rules* for the group of positive examples.

When a new, unlabeled document is presented, we assign its label depending on whether any of the rules are satisfied (i.e., the patterns are found in the document). If all words in any rule are found, the document's label is positive. Otherwise, it is negative.

Figure 3.11 is an example of a set of decision rules induced from a well-known collection of Reuters newswires. The topic is earnings reports. Each rule is a phrase,

Fig. 3.11 Decision rules for
document classification

shr → earn
div → earn
dividend → earn
payout → earn
qtr → earn
earnings & sees → earn
quarter & cts → earn
split → earn
profit → earn
OTHERWISE → ~earn

simply a conjunction of words. If any of these phrases are found, the earnings label
would be assigned as the topic. Otherwise, the last rule gets satisfied and it would
be labeled as not being an earnings report.

In many categorization applications, the categorizer is just a means to assign a
label. The primary objective is to get the label right. For decision rule categorizers, the objective can be broadened. Because the rules are composed of words, and
words have meaning, the rules themselves can be insightful. They can expand our
knowledge and suggest reasons for reaching a conclusion. More than just attempting
to assign a label, such as a topic, to a new document, a set of decision rules induced
from a collection of documents may help summarize how to make decisions. For
example, the rules may suggest a pattern of words found in newswires prior to the
rise of a stock price. Of course, for this example, we would be extremely lucky (and
wealthy!) if we were to find such patterns that were also highly predictive, but it
still drives home the general advantage of rules over other categorizers: patterns of
words have more potential in expanding our base of knowledge and supporting our
decisions than rudimentary scores or measures of similarity. The downside of rules
is that they can be less predictive if the underlying concept is complex. However,
even in these situations, they can lead to insights into the nature of the key predictive
words and phrases.

Although decision rules can be particularly satisfying solutions for text mining,
the procedures for finding them are more complicated than other methods. The expectation is that a relatively small number of words and phrases will provide a good
solution. Yet, the search for these words and phrases that distinguish one class from
the other can be time-consuming and complex. Let's assume that all cells in the
spreadsheet are binary-valued, with each entry indicating the presence or absence
of a word. Because our prediction problem is binary classification, we can readily
apply standard procedures to learn the rules. The primary steps are the following:

1. Find a covering rule set that completely separates the two classes.
2. Prune the rule set into a sequence of smaller rule sets.

   • Optimize each rule set and repeat the pruning process until a single rule re-
     mains.

3. Evaluate the rule sets and pick the best one as the final solution.

Fig. 3.12 Greedy rule
induction for obtaining a
covering rule set

1. Grow conjunctive phrase T (until false positive errors are
   zero) by greedily adding words that minimize error.
2. Record T as the next rule R. Remove documents covered
   by T, and continue with step 1 until all documents are covered.

The covering rule set is almost always an overly optimistic solution in terms of
performance. We try to find rules that completely separate the classes. With a large
dictionary, it may be easy to find very specialized rules that take advantage of small
differences in the training sample of documents. These rules may not hold up in the
future, but they will allow for complete separation of the training documents. No
document in the positive class will be classified in the negative class because no
phrase found for the positive class will be found in the negative class. Figure 3.12
describes a greedy algorithm for learning rules from the sample. We simply keep
adding words to a phrase until no errors are made. The algorithm is greedy because
once it makes a decision on the next word, it does not revise that decision. No
rules are induced for the negative class. If none of the rules apply to an unlabeled
document, its class is negative by default.

These covering rules may appear perfect on the training documents because they
separate the two classes. But a careful reading of the words and phrases may show
that some rules seem to be awkward, almost overspecified. If the learning method
cannot find short phrases to cover lots of documents, the method will substitute
longer phrases that cover fewer documents. If the method keeps adding words, it will
eventually distinguish at least one document from the others. But these rules will
often be too specific, covering few cases and overfitting the training collection of
documents. The rules will not generalize to new documents for correct predictions.
Some of the words may be very good, but less predictive specialized words are
added just to get a rule that excludes the negative class documents.

The decision rules are more than just predictors for new documents. These are
phrases that we can understand, and they should make sense to the reader. Unlike
a distance measure, the decision rules should discriminate among the positive and
negative documents and should also be clear in their rationale. If we see a phrase
that has several reasonable words and then some apparently arbitrary words, we may
question the validity of the phrase.

Because these covering rules and phrases are overspecialized, they will not necessarily be the best for prediction on new examples. It may be better to use simpler
phrases with fewer words. These simplified phrases will not be as perfect as the
full covering set of phrases; they will make mistakes on the training samples. But
more compact rules may have more predictive words and could be more accurate
for predictions on new documents. These simplified phrases can be readily obtained
by pruning the covering set of rules.

How do we determine how much we should simplify a covering set of phrases or
rules? Our rules may be phrases of words, but the process is no different from that
used in numerical applications of machine learning. The phrases are decision rules,
and a large set of rules can be pruned to a smaller set of rules or phrases. Which one

Fig. 3.13 Pruning decision rules

1. Compute err/word for each single deletion operation (word or phrase).
2. Select the operation with the minimum err/word.
3. If more words remain, go back to step 1. Otherwise, the selected set of phrases is the one where err/word is minimum over all cumulative deletion operations.
4. Store the selected set of phrases and repeat the whole process by pruning that set, starting at step 1.

will predict the best? This can be measured empirically without involving human judgment. The original covering rule set will be our most verbose set of words and phrases. Complete words or phrases will be deleted, resulting in a more compact set of rules, the most compact being deleting all words and always choosing the larger class. This concept of pruning is shared by many machine-learning programs. To prune in a controlled manner, a series of smaller and smaller sets of phrases are found, and their predictive performance is measured. The set of phrases with the best predictive performance is selected.

Figure 3.13 is a procedure for pruning a set of phrases. Consider two types of operations: delete a word or a complete phrase. For each operation, compute *err/word*, the number of increased errors per deleted word. We can compute the local err/word for each single operation and the global err/word for the cumulative set of deletion operations, including the current one. For example, if we recursively deleted three phrases, we can compute the increased number of errors per deleted word after all three deletions.

This type of pruning is known as *weakest-link pruning*. It allows us to match different size sets of rules and then pick one set by some standard, usually minimum or near-minimum error. The method prunes rule sets by a complexity measure. The measure that we use is err/word, where we prune a rule set at the point where the number of errors introduced per number of discarded components is minimum. The procedure is repeated on the new, smaller rule set. The result of this process is a series of k rule sets, $RS_1, \ldots, RS_k$, ordered by complexity $C_1, \ldots, C_k$. Each one of these can be evaluated on independent test data, and an error rate can be measured (we will discuss error estimation in Sect. 3.5). An example of this process is illustrated in Table 3.1, where we have seven rule sets. The covering rule set has nine rules or phrases and ten words, and its error is estimated at .1236. Two rule sets are of particular interest. The first of these is the rule set tagged with a '*'. It is the minimum error (as measured by test cases) rule set. A related rule set is the one tagged by '**'. This is the smallest rule set with an error within 1 standard error of the minimum error. Why is this second rule set (which we shall refer to as the *1 SE rule set*) interesting? Ideally, we should accept the rule set with the lowest error (highest predictive power). However, in practice, our estimates of error have an inherent variability. So the performance difference between the minimum-error rule set and 1 SE rule set is not of much significance. Moreover, in the real world, the population characteristics from which our samples are drawn are subject to slight variations over time. The 1 SE rule set, being generally of a lower complexity, performs better in these circumstances. The situation is analogous to buying a pair of shoes: rather

Table 3.1 Example of rule induction error summary

RSet	Rules	Words	TrainErr	TestErr	TestSD	AvgWords	Err/Word
1	9	10	.0000	.1236	.0349	9.9	0.00
2*	6	7	.0337	.1011	.0320	7.0	1.00
3	5	5	.0787	.1236	.0349	5.0	2.00
4	4	4	.0899	.1236	.0349	4.0	4.00
5**	3	3	.1011	.1124	.0335	3.0	1.00
6	2	2	.1910	.1910	.0417	2.0	8.00
7	1	1	.3820	.3820	.0515	1.0	17.00

Fig. 3.14 Optimizing decision rules

```
for every word w in a set of phrases do
    for every dictionary word d_j do
        Compute error of set of phrases with w replaced by d_j
        If this is lower than the original error, replace w by d_j
    endfor
endfor
```

than buying a pair of shoes that fit perfectly at the moment of purchase, one is better off getting a pair with some slight room for future changes.

The phrases are not mutually exclusive. More than one phrase can be found in a document. Pruning the rules by deleting a word can increase overlap, while deleting a phrase may create a hole where some positive documents have no occurrences in any remaining phrases. These issues require an optimization step to be invoked following the pruning of a rule set. One form of optimization is swapping words, also known as backfitting. For any word in our current phrases, we try to replace it with any other word in the dictionary. If another word improves performance, we swap them, moving the new word into the phrase and the old word back into the dictionary. We continue the process until no word can be replaced with another word that reduces errors. Figure 3.14 describes this optimization procedure. The procedure can be implemented very efficiently using dynamic programming techniques. The advantage of backfitting is that it fixes problems in the rule set without changing its size.

3.4.3.1 How to Find the Best Decision Rules

Although decision rule induction is a relatively complex procedure, the interpretability of the result is worth the effort. A solution in the form of words and phrases is compatible with a search engine, and the words can readily be highlighted in the documents. From a training perspective, the answer is intuitive and can be informative and insightful. When compared with k-nearest-neighbor methods, it may seem that one must pay the price in terms of extra learning time for rule induction. However, to get answers of the same predictive quality as rules, the

Fig. 3.15 Rule set selection procedure

> 1. Initialize: $i = 1$; $RS_1 =$ covering rule set;
> 2. Find RS_i with C_i components ($C_i < C_{i-1}$, $i > 1$) and save it;
> 3. If RS_i has more than 1 component, increment i and go to step 2
> 4. Evaluate all RS_i by estimating future error
> 5. Apply criteria to select best rule set:
>   (a) Find min test error
>   (b) Consider all simpler rule sets within one standard error of min test error:
>   (i) Pick the one with highest predictive value, or ask an expert to pick the one that makes sense
>   (ii) Consider only rule sets with predictive value greater than a threshold

k-nearest-neighbor methods may need to be tuned to the right value of k—a process that is nontrivial and requires substantial computational resources.

A data model that employs binary word values is effective and maintains the interpretability of the answers. An alternative is to use a three-value count, 0, 1, 2, where 2 means two or more. The ternary approach is close to binary but is somewhat more flexible in producing compact phrases. A tf-idf representation is incompatible with the interpretive qualities of decision rule learning.

For a binary-valued word representation, all procedures described earlier would try either a positive or negative value. Thus, a phrase could be "word A and NOT word B." The phrases can be restricted to only positive words. This is less flexible than a true binary system, but the rules are more intuitive since they only depend on the presence of words, not their absence. For a ternary representation, $>$ and $<$ operators are also used, so that an answer might be "NOT word A AND occurrences of word B are >1."

Selecting the best rule set leaves some room for human preference. Figure 3.15 describes a procedure for selecting the preferred phrases by using the summary table that accompanies a solution proposed by the learning system. In the absence of any specialized knowledge, we like to always choose the most compact and reasonable set of phrases (say, within one standard error of the minimum, as explained earlier). There is also a possible tradeoff between the number of rules or phrases and the overall size of the rule set (measured by the total number of words). In some circumstances, one may want short phrases (but can tolerate a large number of them); in other circumstances, one may want fewer phrases (but can tolerate long ones). The summary table contains information that allows such preferences to be taken into account in making a selection.

The procedures that have been described for learning the predictive words and phrases are special cases of decision rule induction. We have been able to take advantage of the sparseness of the data. For numerical induction, rule induction methods usually compare all instances of a variable using greater than and less than operations. For text, we know in advance that the values will either be binary or ternary. Based on experience, we favor using local dictionaries of only a few hundred words. The most frequent words are adequate, and the stopwords should be removed. In many benchmarks for text categorization and decision rules, the results of learning with much larger dictionaries do not improve. Some additional speedups may be achieved in the pruning steps. Instead of considering all words in a phrase for pruning, only the last word in a phrase may be examined. The last word in a phrase is typically the most specialized, and the preceding words are often more predictive. The accompanying software includes an implementation of these concepts.

So far, we have seen the similarity measures of the nearest-neighbor methods that require the least effort in learning or assembling a sample, and we have seen decision rules that may take more time to learn but may still be preferred because they are more intuitive and often more accurate. Next, we look at the weighted-scoring methods that have an edge in learning speed or predictive accuracy.

3.4.4 Decision Trees

Decision trees are special decision rules that are organized into a tree structure. A decision tree divides the document space into non-overlapping regions at its leaves, and predictions are made at each leaf. An example decision tree is given in Fig. 3.16.

In this example, each document is represented by two attributes: word A and word B, and we want to classify each document into either category X or category Y. Each node of the tree is a test of the form "attribute value < threshold". For example, the root $A < 2$ implies that we want to check whether word "A" appears less than twice in a document. If so, we move to the left child; otherwise, we move to the right child. Eventually, we reach a leaf node, where a decision is made. One can easily convert a decision tree into a set of non-overlapping decision rules. The most straightforward way to convert a tree into an equivalent set of rules is to create a set with one rule for each leaf by forming the logical conjunction of the tests on the unique path from the root of the tree to the leaf. Thus, reading $\wedge$ as "and," a set of four rules equivalent to the decision tree in Fig. 3.16 is

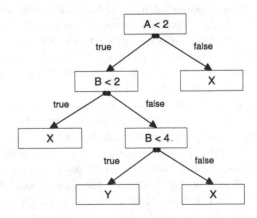

Fig. 3.16 A simple decision tree

$$(A < 2) \wedge (B < 2) \rightarrow X,$$
$$(A < 2) \wedge (B \geq 2) \wedge (B < 4) \rightarrow Y,$$
$$(A < 2) \wedge (B \geq 2) \wedge (B \geq 4) \rightarrow X,$$
$$(A \geq 2) \rightarrow X.$$

Note that the exact form of each test that appears in a rule depends on whether the "true" branch or the "false" branch was taken in going from the root to the leaf. It is possible to simplify the converted rules using more complicated algorithms. However, decision tree itself is a convenient structure that is computationally efficient. For example, using the tree structure, a document can be classified by following a path from the root of the tree to a leave node, which requires no more than L tests, where L is the depth of the tree. On the other hand, the number of leaves, thus the corresponding number of decision rules, may be as many as 2^L.

Decision tree is also easy to train. We can grow the tree greedily node-by-node. At each node, we pick the attribute and threshold so that "attribute value < threshold" gives the best split according to a pre-defined criterion. A general way to view tree splitting is to regard any potential split as the creation of a new feature. Each document reaching the node can take either value 0 or 1 in this "new feature" based on the test result "attribute value < threshold". We can then apply a feature selection method (as described in Chap. 2) to pick the best split (best feature to add).

3.4.5 Scoring by Probabilities

The most obvious method of classification is direct lookup of the probabilities of words in a document. Let C be the class label we are interested in and x be a feature vector that denotes the presence or absence of words from a dictionary. Mathematically, the objective is to estimate $\Pr(C|x)$, the probability of a class, given the presence or absence of words from a dictionary. For singly labeled document collections, we can choose the category C that has the largest probability score $\Pr(C|x)$. For multiply labeled document collections, if our interest is to maximize the accuracy, then C is selected whenever $\Pr(C|x)$ is greater than 0.5. Another way to look at the multiply labeled case is that for each label we divide the document collection into two classes: one class with label C and the other class with a label that is not C. Therefore, we have a binary classification problem for each label value C. This is reasonable because the multiple labels assigned to documents are usually independent of each other, and hence it is possible to view each label assignment as a separate classification problem with two classes (labeled and not labeled). It thus suffices to consider the binary class problem.

However, we know that, even for this problem, a complete computation of probability is impossible. Even a 100 word dictionary has 2^{100} possible combinations. Still, a simplified approach to probability estimation, called *Bayes with independence* or *naive Bayes*, has often been attempted. The mathematics is straightforward and the computation is efficient, which leads to wide application of this approach, especially in applications where a quick implementation takes priority over accuracy.

$$Pr(C|x) = Pr(x|C) * Pr(C)/Pr(x). \tag{3.2}$$

Bayes' rule is given in (3.2), where C is the class of interest and x is a vector of ones and zeros corresponding to the presence or absence of dictionary words for a specific document.

Note that when applying (3.2) for two or more classes, the common factor of $1/Pr(x)$ does not change the relative ranking of $Pr(C|x)$. Therefore, it does not need to be computed explicitly for ranking purposes. If it is dropped from the comparisons, one does not have an explicit probability estimate, only the numerator. However, it might be useful to compute $Pr(x)$ since it allows one to compute a probability estimate. Using probability estimates, one can adjust the kinds of errors made and also specify a *reject probability threshold* that must be exceeded to classify. When there are two classes, C_1 and C_2, $Pr(x)$ is readily computed as in (3.3).

$$Pr(x) = Pr(x|C_1) Pr(C_1) + Pr(x|C_2) Pr(C_2). \tag{3.3}$$

The key to using these equations is to compute $Pr(x|C)$. If we assume that the words are independent then instead of looking up the probability of the complete vector of x, we can look up the probability of the presence or absence of each word, $Pr(x_j|C)$, and multiply them all together. We use x_j to denote the j-th component of x. Equation (3.4) states this mathematically.

$$Pr(x|C) = \prod_j Pr(x_j|C), \qquad Pr(x) = \sum_C Pr(C) \prod_j Pr(x_j|C). \tag{3.4}$$

The conditional probabilities in (3.4) are readily estimated if one uses the simple binary presence or absence of a word as a feature value that would give only two possible values for each feature.

The probability estimates are easy to obtain from our spreadsheet. $Pr(C)$ is determined from the frequency of ones in the last column divided by n, the number of examples, $freq(C)/n$. Each x_j is either a 1 or a 0 (presence or absence of the word w_j). The quantity $Pr(x_j = 1|C)$ is computed from the frequency of ones for x_j, where only the examples labeled C are considered, $freq(x_j = 1, label = C)/freq(C)$. The probability of w_j not occurring in C, $Pr(x_j = 0|C)$, is $1 - Pr(x_j = 1|C)$.

Figure 3.17 is an example for a dictionary of four words. The training sample consists of ten documents of which four are labeled as Class $= 1$ and the remaining six as Class $= 0$. We can easily compute estimates of the various conditional probabilities as shown in the figure. Now suppose we get a new document D that has w2, w3, and w4. Then, for the positive class, we could compute D's probability as

$$Pr(Class = 1|D) = ((1 - .75) * .25 * .5 * .5) * .4/Pr(D) = .00625/Pr(D).$$

For the negative class, the probability would be computed as

$$Pr(Class = 0|D) = ((1 - .5) * .67 * .33 * .5) * .6/Pr(D) = .03333/Pr(D),$$

w1	w2	w3	w4	Class
1	0	0	1	1
0	0	0	1	0
1	1	0	1	0
1	0	1	1	1
0	1	1	0	0
1	0	0	0	0
1	0	1	0	1
0	1	0	0	1
0	1	0	1	0
1	1	1	0	0

	Class=1	Class=0
Pr(Class)	0.40	0.60
Pr(w1 | Class)	0.75	0.50
Pr(w2 | Class)	0.25	0.67
Pr(w3 | Class)	0.50	0.33
Pr(w4 | Class)	0.50	0.50

Fig. 3.17 Scoring by probabilities

and as a result the document D would be labeled as Class $= 0$ (if one computes $\Pr(D)$, one gets a probability of 0.84 for the classification).

The performance on text benchmark applications for naive Bayes is usually weaker than for the other methods described in this chapter. Still, it requires almost no memory and little computation, so it does have its advocates. It usually works best with a relatively small dictionary representing the key words needed to make a decision for that class. An implementation is provided in the accompanying software.

The naive Bayes method of estimating probabilities looks complex, but in fact it has a linear structure. This can be seen by noting that, given a binary feature vector x, the probability score of class C is

$$\Pr(C|x) = \frac{\Pr(C) \prod_j \Pr(x_j = 0|C)}{\Pr(x)} \prod_j \left(\frac{\Pr(x_j = 1|C)}{\Pr(x_j = 0|C)}\right)^{x_j},$$

which can be rewritten as

$$\Pr(C|x) = \frac{1}{\Pr(x)} \exp\left(\sum_j w_j x_j + b\right), \tag{3.5}$$

where

$$w_j = \ln \frac{\Pr(x_j = 1|C)}{\Pr(x_j = 0|C)}, \qquad b = \ln \Pr(C) + \sum_j \ln \Pr(x_j = 0|C). \tag{3.6}$$

This formulation is often called the *multivariate Bernoulli model*. Another naive Bayes model, referred to as the *multinomial model*, replaces the linear weights formula in (3.6) with the one in (3.7). Here n is the number of examples and m is the number of features.

$$w_j = \ln \frac{\lambda + \text{freq}(x_j = 1, \text{label} = C)}{\lambda m + \sum_{j'=1}^{m} \text{freq}(x_{j'} = 1, \text{label} = C)}, \qquad b = \ln \frac{\text{freq}(\text{label} = C)}{n}. \tag{3.7}$$

The multinomial model is frequently used in text categorization applications. It normalizes the length of a document, which often leads to slightly better performance because it is less sensitive to the effect of document length than the multivariate Bernoulli model. The parameter $\lambda > 0$ is a smoothing parameter, often set to 1 in the literature. However, we find that a smaller value such as 0.01 can sometimes be more effective.

In these forms, it is easier to see that one might also use other methods to directly train the linear weights. We will examine linear scoring methods in the next section.

3.4.6 Linear Scoring Methods

In order to achieve good prediction performance, it is often necessary to create a feature vector of very high dimension. Although many of the features are not useful, it can be difficult for a human to tell what feature is useful and what feature is not. Therefore, the prediction algorithm should have the ability to take a large set of features and then select only useful features from the full set. A very useful method to achieve this is by using linear scoring.

The naive Bayes method described above can be regarded as a special case of the linear scoring method. This can be seen clearly from (3.5). However, the performance can be significantly improved using more sophisticated training methods to obtain the weight vector $w = [w_j]$ and bias b.

Consider the problem of distinguishing between two classes. The general scoring method is to assign a positive score to predict the positive class and a negative score to predict the negative class. Figure 3.18 illustrates an example of using a set of weights to determine the score for a document. For all words that occur in a document, we find their corresponding weights. These weights are then summed to determine the document's score.

Mathematically, this method is a linear scoring function. The general form is in (3.8), where D is the document and w_j is the weight for the j-th word in the dictionary, b is a constant, and x_j is a one or zero, depending on the j-th word's presence or absence in the document.

$$\text{Score}(D) = \sum_j w_j x_j + b = w \cdot x + b. \qquad (3.8)$$

Linear scoring methods are classical approaches to solving a prediction problem. The weaknesses of this method are well-known. Geometrically, the method can be described as producing a line or hyperplane. Although a line cannot fit complex surfaces, and a curvy shape might be needed, it is often possible to create appropriate non-linear features so that a curve in the original space lies in a hyperplane in the enlarged space with the additional nonlinear features. In this way, nonlinearity can be explicitly captured by constructing sophisticated nonlinear features. An advantage of this approach is that the modeling aspect becomes conceptually very simple

Fig. 3.18 Computing the weighted score of a document

Linear Model

Word	Weight
dividend	0.8741
earnings	0.4988
eight	−0.0866
extraordinary	−0.0267
months	−0.1801
payout	0.6141
rose	−0.0253
split	0.9050
york	−0.1908
...	...

New Document

Words	Score
dividend, payout, rose	1.4629

since we can focus on creating useful features and let the learning algorithm determine how to assign a weight to each feature we create. Another advantage is that the linear scoring method can efficiently handle sparse data. This is important for text-mining applications since although feature vectors can have high dimensionality, they are usually very sparse.

We know from various benchmarks that the linear scoring approach does surprisingly well on text classification, to the point where it rivals the best results on benchmark data. Text lends itself to a scoring approach such as the one we described in Fig. 3.18.

The modern approach to learning the weights is not the same as the classical statistical methods. The simple naive Bayes methods have severe problems with redundant attributes, which in text corresponds to words that behave like synonyms. Classical methods were developed to handle a small number of attributes, certainly not the tens of thousands of words in a global dictionary. The newer linear methods are oblivious to these limitations. A major advance in linear methods for text has been their ability to work with huge dictionaries and find weights for every word in a complete dictionary. If there are ten synonyms, it can weigh each one. This capability to work with so many words and weigh all of them both positively and negatively seems to capture the subtleties of language, where some words are precise and strong predictors and others are vague and weak predictors.

Surely, isn't the computational time for learning in this high-dimensional space prohibitive? Not at all. The same problem that cannot be solved by a classical method can now be solved incredibly quickly, much faster than by nearest-neighbors similarity methods or the decision rules. Moreover, the natural extension for the linear model is not to find more complex mathematical functions. Instead, scoring might be extended by adding word phrases to the single-word dictionaries. Benchmarks using more complex scoring methods generally perform no better than the linear scores.

The key problem with these weighted-scoring methods is that of learning the weights, the second column in Fig. 3.18. The words are those in the dictionary, and the weights for them will be learned from a collection of documents. The label is assigned by applying (3.8). How do we learn the weights? Implementations can consist of less than 200 lines of code. However, the method is a mathematical process, an application of numerical analysis.

3.4.6.1 How to Find the Best Scoring Model

So how do we learn the weights? To determine the most efficient way, the treatment is necessarily mathematical. It may look complex, but the implementation in software is straightforward. The representation of the words can be binary, but the tf-idf transformation usually yields better results.

Let us first look at the mathematics behind the procedure. We consider a two-class prediction problem to be one that determines a label $y \in \{-1, 1\}$ from an associated vector x of input variables. Given a continuous model $p(x)$, we consider the following prediction rule: predict $y = 1$ if $p(x) \geq 0$, and predict $y = -1$ otherwise. The classification error (we ignore the point $p(x) = 0$, which is assumed to occur rarely) is

$$I(p(x), y) = \begin{cases} 1 & \text{if } p(x)y \leq 0, \\ 0 & \text{if } p(x)y > 0. \end{cases}$$

A useful method for solving this problem is by linear predictors. These consist of linear combinations of the input variables $p(x) = w \cdot x + b$, where w is often referred to as *weight* and b as *bias*. We call (w, b) the weight vector and use the term bias for statistical bias.

Let (x^i, y^i) be the i-th row of the spreadsheet, where x^i is the vector representation of the i-th training data, and y^i represent the label, which takes the value 1 if the document belongs to category C and value -1 otherwise. Note that, for notation simplicity, we have slightly changed the representation in our spreadsheet model (using -1 instead of 0 to represent the label of negative data). A very natural way to compute a linear classifier is by finding a weight $(\hat{w}, \hat{b})$ that minimizes the average classification error in the training set:

$$(\hat{w}, \hat{b}) = \arg\min_{w,b} \frac{1}{n} \sum_{i=1}^{n} I(w \cdot x^i + b, y^i). \tag{3.9}$$

Unfortunately, this formulation leads to a nonconvex optimization problem which may have many local minima. Finding the global optimal solution, which we desire, is generally very hard (to be mathematically precise, it is *NP-hard*). It is thus desirable to replace the nonconvex classification error loss $I(p, y)$ with a convex formulation that is computationally more desirable.

In order to motivate a convex formulation, we shall first consider the situation where the problem is linearly separable. In this case, we want to find a linear separator (w, b) such that $w \cdot x^i + b < 0$ when $y^i < 0$, and $w \cdot x^i + b > 0$ when $y^i > 0$. That is, we want to find w such that $(w \cdot x^i + b)y^i > 0$ for all i. Note that if a linear separator w exists, then there are more than one linear separators, since a small perturbation of w still separate the data. It is thus preferable to find an "optimal" linear separator. An idea, popularized by Vapnik, is to find the most stable linear separator, so that any small perturbation of x^i does not change the classification rule. The stability of a linear separator's prediction on the i-th point (x^i, y^i) can be measured by its margin: $\gamma^i = (w \cdot x^i + b)y^i / \|w\|$, where $\|w\|^2 = w \cdot w = \sum_j w_j^2$.

Fig. 3.19 Margin and linear separating hyperplane

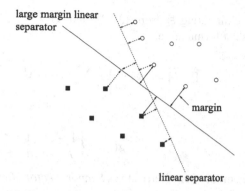

large margin linear separator

margin

linear separator

If the margin γ^i is large, then the prediction on this data point is stable. In order to change the sign of the prediction, we have to move $x^i \rightarrow x^i + \Delta x^i$ such that $(w \cdot (x^i + \Delta x^i) + b)y^i < 0$. That is, $-w \cdot \Delta x^i y^i / \|w\| > \gamma^i$. The larger γ^i is, the larger modification Δx^i is needed to switch the sign.

The margin idea is illustrated in Fig. 3.19. We want to find a linear separator such that the smallest γ^i is as large as possible. Mathematically, the optimization problem for finding the linear separator with the largest margin is

$$\min_{i=1,\ldots,n} (w \cdot x^i + b)y^i / \|w\|.$$

A more popular method, which is equivalent to the above formulation, is to minimize $\|w\|$ under the constraint $\min_i (w \cdot x^i + b)y^i \geq 1$. That is, the optimal hyperplane is the solution to

$$[\hat{w}, \hat{b}] = \arg\min_{w,b} \|w\|^2$$

$$\text{subject to} \quad (w \cdot x^i + b)y^i \geq 1 \quad (i = 1, \ldots, n).$$

If the data is not linearly separable, then the idea of margin maximization cannot be directly applied. Instead, one considers the so-called soft-margin formulation as follows:

$$[\hat{w}, \hat{b}] = \arg\min_{w,b} \left[\|w\|^2 + C \sum_{i=1}^{n} \xi^i \right]$$

$$\text{subject to} \quad y^i (w \cdot x^i + b) \geq 1 - \xi^i, \quad \xi^i \geq 0 \ (i = 1, \ldots, n). \tag{3.10}$$

The parameter ξ^i are variables that allows the margin constraint for each point i to be violated (when $\xi^i > 0$). However, a linear penalty is included into the optimization when margin constraint is violated. The larger C is, the heavier the penalty of such violation. In particular, if $C \rightarrow \infty$, we obtain back the original separable margin maximization formulation.

By eliminating ξ_i from (3.10), and let $\lambda = 1/(nC)$, we obtain the following equivalent formulation:

$$[\hat{w}, \hat{b}] = \arg\min_{w,b} \left[\frac{1}{n} \sum_{i=1}^{n} g((w \cdot x^i + b)y^i) + \lambda \|w\|^2 \right], \tag{3.11}$$

where

$$g(z) = \begin{cases} 1 - z & \text{if } z \leq 1, \\ 0 & \text{if } z > 0. \end{cases} \tag{3.12}$$

This method, often referred to as *Support Vector Machine* (SVM), can be regarded as a modification of the classification error minimization formulation (3.9), where we replace classification error minimization by minimizing the following upper bound:

$$\min_{w,b} \frac{1}{n} \sum_{i=1}^{n} g(w \cdot x^i + b, y^i).$$

The additional term $\lambda \|w\|^2$ is called regularization, which we will discuss later on.

The loss function $g(\cdot)$ used in SVM is often referred to as *hinge loss*. The resulting optimization problem is described as *convex* and thus computationally tractable. This modification has been popular with many practitioners. However, we consider a different method that minimizes the following loss function:

$$h(p, y) = \begin{cases} -2py, & py < -1, \\ \frac{1}{2}(py - 1)^2, & py \in [-1, 1], \\ 0, & py > 1. \end{cases}$$

This loss function has its root in the robust regression literature, and we call it *robust classification loss*. Our linear weights are computed by minimizing the following average loss on the training data:

$$(\hat{w}, \hat{b}) = \arg\min_{w,b} \frac{1}{n} \sum_{i=1}^{n} h(w \cdot x^i + b, y^i). \tag{3.13}$$

The main advantage of using (3.13), (3.14), or (3.15), is that it is possible to show that one computes a weight $(\hat{w}, \hat{b})$ so that the conditional in-class probability $\Pr(y = 1|x)$ can be estimated as $\hat{q}(x) = \max(0, \min(1, (\hat{w} \cdot x + \hat{b} + 1)/2))$. The quantity $\hat{q}(x)$ can be interpreted as an estimate of the statistical confidence of the classifier's prediction, which is useful to know in many problems.

A problem of using (3.13) without regularization is that its solution may not be unique. Such a formulation is often called *numerically ill-posed*. It is difficult to design good numerical algorithms for solving ill-posed problems since a small perturbation of the formulation can cause a large instability in the solution. In general, (3.13) may be ill-posed and numerically unstable when the dimension of x is larger than the number of training examples n, which frequently happens in text-mining applications. A closely related statistical problem is that using (3.13) directly to compute a linear classifier can cause overfitting. A common remedy to the problems above is to restrict the search space of linear classifiers. This method is often called *regularization*, which avoids overfitting. Regularization can be achieved in many different ways. For example, we may consider searching over the weight space defined as $\|w\|^2 \le A$, which introduces a constraint (or regularization) on w. Here, we consider a slight different constraint which includes a constraint on b in addition to w:

$$\|w\|^2 + b^2 \le A.$$

A is a parameter controlling the size of the search space. This method is quite popular. The resulting method of computing a linear weight can now be written as

$$(\hat{w}, \hat{b}) = \arg\min_{w,b} \frac{1}{n} \sum_{i=1}^{n} h(w \cdot x^i, y^i), \qquad \|w\|^2 + b^2 \le A. \qquad (3.14)$$

From the computational point of view, one can introduce a Lagrangian multiplier λ for the constraint $w^2 + b^2 \le A$ and obtain the following equivalent formulation:

$$(\hat{w}, \hat{b}) = \arg\min_{w,b} \left[\frac{1}{n} \sum_{i=1}^{n} h(w \cdot x^i + b, y^i) + \frac{\lambda}{2}(\|w\|^2 + b^2) \right]. \qquad (3.15)$$

This formulation resembles (3.11), and will be what we use in our computation. Because it minimizes the averaged robust classification loss function, we call it *robust risk minimization* (RRM). The advantage of (3.15) is that it transforms a constrained optimization problem into an unconstrained optimization problem, which is typically easier to handle numerically. In this transformation, the regularization parameter A is replaced by λ, which in general can be tuned using test data as described in Sect. 3.5. Other loss functions may be used as well. Figure 3.20 shows the robust loss and hinge loss functions against the classification error function.

In order to handle large data, it is necessary to use an algorithm that can take advantage of the sparse structure in the word vector representation of documents. Two types of algorithms can be proposed. One works by going through each word in the dictionary and updating the corresponding weight component; the other works by going through each data point (i.e., each positive word in a document) and updating the weight accordingly. The second approach is closely related to a class of learning methods called *online learning* that is desirable for large problems. Since such an algorithm examines the data sequentially, it does not even need to store all the training data in memory at the same time. This allows the algorithm to handle a large amount of data without potential memory issues.

For large problems such as document collections, we use the online-style method to solve (3.15). This involves a transformation of (3.15) into an equivalent dual

Fig. 3.20 Plots of some loss
functions

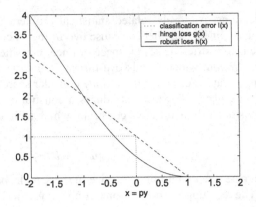

formulation that depends on a set of variables $\alpha = [\alpha_i]$, where each variable α_i
$(i = 1, \ldots, n)$ is associated with a data point (x^i, y^i). In particular, the solution of
(3.15) has the dual representation

$$\hat{w} = \sum_{i=1^n} \hat{\alpha}_i x^i y^i, \qquad \hat{b} = \sum_{i=1^n} \hat{\alpha}_i y^i$$

where $\hat{\alpha}$ is the solution to the following dual optimization problem:

$$\hat{\alpha} = \arg\min_{\alpha} \left[\sum_{i=1}^{n} \left(\frac{\lambda n}{2} \alpha_i^2 - \alpha_i \right) + \frac{1}{2} \left\| \sum_{i=1}^{n} \alpha_i x^i y^i \right\|^2 + \frac{1}{2} \left(\sum_{i=1}^{n} \alpha_i y^i \right)^2 \right]$$

$$\text{s.t.} \quad \forall i : \alpha_i y^i \in \left[0, \frac{2}{\lambda n} \right]. \tag{3.16}$$

The equivalence follows from a general convex-duality formalism, which is beyond the scope of this book. We use an alternating direction optimization method to
solve (3.16). The idea is to cycle through all data points (x^i, y^i) $i = 1, \ldots, n$, where
for each i we update the corresponding dual variable α_i so as to decrease the objective value in (3.16), while keeping the remaining dual variables α_j fixed ($j \neq i$).
Each step is thus a simple one-dimensional optimization problem with respect to α_i,
which can be minimized exactly in our case. However, instead of using the exact optimization at each step, we employ a gradient descent rule, which is more generally
applicable:

$$\alpha_i \rightarrow \alpha_i - \eta_i [\lambda n \alpha_i - y^i - (w \cdot x^i + b) y^i],$$

$$\text{where } w = \sum_{i=1}^{n} \alpha_i x^i y^i, \ b = \sum_{i=1}^{n} \alpha_i y^i. \tag{3.17}$$

In the formula above, η_i is a small step-size quantity that is sometimes referred to
as the "learning rate" in the online learning literature. In addition, since we have
the constraint $\alpha_i \in [0, 2/\lambda n]$ in (3.16), we choose η_i to ensure that the updated α_i

Input: training data $(x^1, y^1), \ldots, (x^n, y^n)$
Parameters: $K, c, \eta_1, \ldots, \eta_n$
Output: weight vector w_j $(j = 1, \ldots, m), b$
Initialize: $\alpha_i = 0$ $(i = 1, \ldots, n)$; $w_j = 0$ $(j = 1, \ldots, m)$; $b = 0$
for $k = 1$ to K **do**
    for $i = 1$ to n **do**
        $p = (w \cdot x^i + b) y^i$
        $d_i = \max(\min(2c - \alpha_i, \eta_i((c - \alpha_i)/c - p)), -\alpha_i)$
        $w = w + d_i x^i y^i$
        $b = b + d_i y^i$
        $\alpha_i = \alpha_i + d_i$
    endfor
endfor

Fig. 3.21 Learning the weights for the linear categorizer

satisfies this constraint (which is also equivalent to a truncation of the updated α_i onto $[0, 2/\lambda n]$). The quantity $[\lambda n \alpha_i - y^i - (w \cdot x^i + b) y^i]$ is the gradient of the dual objective function of (3.16) with respect to α_i, and thus the gradient descent update above modifies α_i in the direction that decreases the objective function. In our case, there is a closed-form solution of η_i such that the update (3.17) exactly minimizes (3.16) as a function of α_i with α_j $(j \neq i)$ fixed. However, choosing a smaller step size η_i usually helps. Moreover, good performance can be achieved by simply picking a small fixed step size η_i, where $\eta_i = 0.25$ is a reasonable generic choice.

We have shown the mathematics behind the procedure for learning weights to show the crucial reformulations that enable an efficient iterative algorithm to be derived. The complete algorithm for learning the weights is shown in Fig. 3.21. A version of it has been implemented and included in the accompanying software. The algorithm is scalable and can efficiently handle very large problems.

In the algorithm above, $c = 1/\lambda n$. The algorithm can be terminated when a certain stopping criterion is met. For example, one criterion is the so-called *duality gap*, which is frequently used in the optimization literature. However, a simpler method of using a fixed number of iterations K generally works well. For simplicity, we often use a fixed number $K = 40$. In our implementation, we also employ a random data ordering of (x_i, y_i). This is to avoid the situation that data in the same category (or similar data) are grouped together. In such a case, an online type algorithm is likely to perform poorly since it will adjust the weight vector to overfit one category before it sees other categories.

Notice that a smaller λ in (3.15) corresponds to a larger A in (3.14). This means that with a smaller λ, a larger weight space is searched, and thus the statistical bias is smaller in the sense that we can approximate the target function more accurately. However, with a large A, statistical variance is larger in the sense that one can also approximate a noisy function more accurately, which causes overfitting of the training data. Conversely, a larger λ means searching a smaller weight space and is less likely to overfit the training data; however, since the model space is smaller, we have a larger bias. It follows that an appropriate choice of λ is necessary for optimal performance. It has been suggested that the choice of $\lambda = 10^{-3}$ is typically good for many text data sets. In our implementation, we use this value as the default. However, for some problems, other values (such as $\lambda = 10^{-4}$) are better. In particular, this is true for large data since it is less likely to overfit. This means that one can search over a larger weight space to reduce the bias. This improves the performance when the training data size becomes large.

Methods to obtain linear scoring functions presented here are designed for binary classification problems. For multi-category problems with more than two classes, we may train separate scoring functions for different categories (each scoring function predicts whether the class label is one of the categories). We can then pick the class label with the largest score as the predicted category. For example, if we want to classify a text document x into three categories "finance", "religion", or "sport", we may train three linear scoring functions: $f_{fin}(x)$ indicating whether x belongs to category "finance", $f_{rel}(x)$ indicating whether x belongs to category "religion", and $f_{spo}(x)$ indicating whether x belongs to category "sport". We then pick the category with the largest score as the winner: if $f_{fin}(x) > f_{rel}(x)$ and $f_{fin}(x) > f_{spo}(x)$, then we classify x into category "finance". This method, usually referred to as the one-versus-all or the winner-takes-all approach, is a standard approach to solve multi-category problems.

3.5 Evaluation of Performance

The learning methods provide potential solutions. They do not guarantee that these are good solutions. To get the best results, we must ensure that (a) the wisest choices are made in applying the methods and (b) estimates are found for the future performance of proposed solutions. Let's look at the immediate task of evaluating a solution and estimating its future performance.

3.5.1 Estimating Current and Future Performance

The standard statistical model assumes that a sample is randomly drawn from some general population as described in Fig. 3.22. The new examples are unlabeled, and

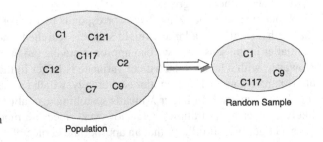

Fig. 3.22 Drawing a random sample from a population

Fig. 3.23 Partitioning documents into training and test sets

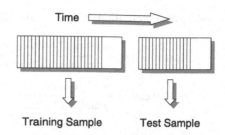

Time

Training Sample   Test Sample

their labels will be assigned. To evaluate the performance of a solution, we train on one sample and test on another sample. Typically, our data might be randomly divided into two parts: one for training and one for testing.

Are these new documents from the same population? Very often they are not, but over relatively short time horizons, we assume that new documents are similar to old documents or that events will unfold in a similar way. Documents come with a time stamp. Almost all evaluation of text-mining solutions orders a sample by time and uses the earlier documents for training and the later documents for testing. This is illustrated in Fig. 3.23, where the sample is divided by time, not randomly. This breakdown more closely simulates the prediction of future events. Unlike the classical random sample for identically and independently distributed data, some additional experimentation may be necessary to identify the time periods that are best for training so that they continue to hold over some fixed future time period.

Once the data are split into training and testing samples, learning takes place exclusively on the training set. Performance can be estimated in terms of several measures. The standard measure for classification is the error rate of (3.18), and its standard error is given in (3.19). The error rate is binomially distributed and is approximately normal. Two standard errors are often used to approximate 95% confidence bounds.

$$\text{Error rate}(erate) = \frac{\text{number of errors}}{\text{number of documents}}, \tag{3.18}$$

$$\text{Standard Error}(SE) = \sqrt{\frac{erate * (1 - erate)}{\text{number of documents}}}. \tag{3.19}$$

Although error-rates and associated standard errors are useful for estimating the performance of predictors in general, for most text applications, such as text categorization, a more detailed analysis of the errors is desirable. For information retrieval applications, there is usually a large number of negative data. A classifier can achieve a very high accuracy (i.e., a very low error rate) by simply saying that all data are negative. It is thus useful to measure the classification performance by ignoring correctly predicted negative data and then examining the sorts of errors made by the classifier. Three ratios have achieved particular prominence: precision, recall, and F-measure. Their definitions are given in (3.20).

$$
\text{precision} = \frac{\text{number of correct positive predictions}}{\text{number of positive predictions}},
$$

$$
\text{recall} = \frac{\text{number of correct positive predictions}}{\text{number of positive class documents}}, \qquad (3.20)
$$

$$
F\text{-measure} = \frac{2}{1/\text{precision} + 1/\text{recall}}.
$$

Precision, recall, and their combination in the F-measure are all more interesting measures of the quality of binary decisions on documents.

The following example illustrates these measures of performance. Let's assume that there is a database of labeled documents. Let's focus on a particular label, such as *sports*. Now consider a classifier that labels documents as *sports* or not, and let's use it to retrieve all the documents that it labels. We can assess the performance of the classifier from the set of retrieved documents by computing the three measures as follows:

- The percentage of all sports documents that are retrieved is the recall.
- The percentage of documents that it correctly labels as sports is the precision.
- F-measure is defined as the harmonic mean of precision and recall. It is often used to measure the performance of a system when a single number is preferred.

Because document collections are typically large, high precision is often more valued. For high precision, the computer's positive decisions are usually correct, but it may fail to catch all positives (this is measured by recall). Thus, if a program identifies spam e-mail with high precision and low recall, it may often leave spam in your Inbox (low recall), but when it puts a spam document in the trash, it is usually correct (high precision).

Is it possible to adjust the precision and recall of a classifier? Since precision and recall measure different kinds of errors, if the overall error rate remains the same, increasing the precision (reducing one kind of error) lowers the recall (increases the other kind of error). This leads to a *precision–recall tradeoff*. Most classifiers have a way of making this tradeoff by a simple variation of a constant. For the classifiers discussed in this chapter, the process is as follows:

- For k nearest-neighbor methods, the threshold can be varied from a simple majority to some other value. For example, if five nearest neighbors being used classify a document within a topic, instead of requiring that three of the five nearest neighbors belong to the topic, one may change this threshold to a different value. A value less than 3 would boost recall, whereas values greater than 3 would boost precision.
- For decision rules, the cost of different kinds of errors can be altered. For example, if false negative errors are made twice as costly as false positive errors, then recall would be boosted.

- For probabilistic scoring, the threshold for a class can be altered from 0.5 to some other value. Lower thresholds would boost recall, while higher values would boost precision.
- For linear models, the threshold can be changed from zero to a different value. Lower values would help recall, and higher values would boost precision.

3.5.2 Getting the Most from a Learning Method

These measures tell us how well we might project into the future. Because the sample may change over time, we can expect somewhat worse results in the future. Evaluation and estimation of future performance is an important objective. The process that is used for documents almost always employs a distinct time-oriented train and test sample. During training, we have another objective. We must set the parameters of our learning program to get the best solutions. Each of the learning methods that we have described has a single parameter related to the complexity of the solution. This parameter can be adjusted during training. For the k-nearest-neighbor methods that compute similarity, the value of k must be determined. For the decision rules, the total number of words in a rule set is typically used as the complexity measure. For the linear model, a regularization parameter controls complexity. How can these methods find the best value during training? For this purpose, the classical approach is effective, where the training data are randomly divided and some of the data are used to evaluate the best value of the complexity parameter. For a very large sample, as is typical in text mining, the risk is not great for tuning a single complexity parameter directly on the test data. The estimate for future performance is usually the basis for selecting the best setting of the complexity parameter. If the same test data are used for selection and estimation, then the estimate may be somewhat optimistic. The safest procedure is to use two samples, one for tuning and one for estimation.

3.6 Applications

The prototypical predictive text-mining application has been text categorization. Newswires are automatically assigned topics such as financial or sports stories. A more visible application has emerged in the public consciousness: e-mail spam. In its simplest form, filtering valid e-mail from spam is an instance of binary classification: spam or not spam. Most e-mail programs allow a user to specify filters. It comes as no surprise that the filters use user-defined decision rules. These are immediately interpretable, and users are very comfortable specifying their own rules composed of words in various heading fields or the body of the message. Some specialized filtering programs at the system level are partially precompiled linear scoring functions, such as noted in our table of words and weights. They allow the administrator to adjust the weights for phrases or other characteristics of e-mail such as the number of spaces in the subject line. E-mail programs may also learn directly from examples, where users mark the spam e-mail, separating all e-mail in an Inbox into spam and not spam. Typically, a linear scoring function is then learned to replace the user in marking incoming e-mail. Obviously, it is more dangerous to move a good message into the trash than not to detect some spam e-mail. Precision takes priority over recall.

We will have more to say about applications of prediction in Chap. 8. Compared with mining of structured data, predictive text mining is a new frontier. Because text mining involves the transformation of unstructured information into structured information, we can expect many applications that mix both numerical and text data.

3.7 Summary

Once text is transformed into numerical vectors, automated prediction methods can be applied. Prediction from text is described in terms of an empirical analysis that can be related to word patterns, especially for document classification. This chapter reviews what should be expected for prediction and the minimum necessary numbers of documents. Fundamental methods of learning from sample data are outlined including similarity-based methods, decision rules and trees, probabilistic methods and linear methods. Evaluation techniques are examined to estimate future performance and to maximize empirical results.

3.8 Historical and Bibliographical Remarks

The predictive methods discussed above have been used in many other contexts. Our interest, however, is their use in text processing and text mining.

The use of nearest-neighbor methods in document retrieval was discussed in the first SIGIR conference (Eastman and Weiss 1978), where the document collection was organized as a tree. Another entirely different example is described in Masand *et al.* (1992), in which the collection is stored in a highly parallel processor, the Connection Machine. In this example, the documents are news stories that are to be assigned classification codes for the Dow Jones news service. Rule-based methods were applied to document classification in Apté *et al.* (1994) using inductive learning to find classification rules. A somewhat different rule system for classifying e-mail, the Ripper algorithm, is described in Cohen (1996). Rule-based methods are actively explored in Maloof (2003), which discusses learning concepts that change over time.

Linear classification methods have become very popular, due both to their simplicity and their good classification accuracy. The naive Bayes method has been widely used in machine learning due to its simplicity. One of the first applications of this method to text categorization was described in McCallum and Nigam (1998), where the authors proposed the two naive Bayes models we discussed in this chapter.

In addition to the standard naive Bayes algorithms, one can also use a simpler linear classification method called the *centroid method*. The linear weight vector for a category is simply the mean of document vectors in the category. Statistically, it can be regarded as a naive Bayes method under the assumption that each vector component is generated according to a Gaussian distribution. Because the Gaussian assumption is not very suitable for text data, this method is often inferior to more standard naive Bayes models. However, it can still be useful for special purposes. Since the method is widely used in text clustering, we discuss it in Sect. 5.2.1.1.

Another classical linear classification method is logistic regression, which is closely related to the maximum entropy (MaxEnt) method. We delay the description of MaxEnt to Chap. 6 since it is one of the main learning methods used in natural language processing and information extraction.

The current popularity of linear classification is mostly due to the development of large-margin classification, generally regarded as a very significant advance of machine-learning methodology. This class of methods includes many state-of-the-art learning algorithms such as boosting and support vector machines (SVM). Generally speaking, a large-margin algorithm finds a classifier that minimizes a certain convex upper bound of the classification error loss.

The idea of boosting can be used to produce a strong learner by combining weaker learners obtained on reweighted samples of the training data (Freund and Schapire 1997). One can also view boosting as greedily minimizing a convex loss of the combined weak learners (Breiman 1999; Schapire and Singer 1999; Friedman *et al.* 2000; Hastie *et al.* 2001). The method has been applied to text categorization with great results. In particular, a version of boosted multiple decision trees was used in Weiss *et al.* (1999) to produce one of the best reported text categorization performances on the standard Reuters data.

Another very popular large-margin method related to boosting is SVM, proposed by Vapnik and based on some theoretical and algorithmic considerations (Vapnik 1998). This method has also been applied to text categorization with very good performance (Joachims 1998). SVM is a specific case of a family of algorithms that we refer to as *regularized linear classification methods*. It also includes the robust risk minimization method (equation (3.15)) described in this chapter. In general, a regularized linear classification method picks a convex loss function (the h-term in (3.15)) to minimize and includes a regularization term (the $\|w\|^2 + b^2$ term in (3.15)) to stabilize the solution. Compared with SVM, the advantage of (3.15) is that one can obtain probability information from its linear scoring output. In principle, this is not possible with SVM. We refer readers to Zhang (2004) for a study on some theoretical properties of different loss functions.

The duality formalism leading to the linear scoring algorithm in this chapter can be found in Zhang (2002). The algorithm that we presented here first appeared in Damerau *et al.* (2004). Although we only described one particular loss function and one particular regularization condition, other versions of regularized linear classifiers can also perform well for text categorization (Zhang and Oles 2001). Moreover, a recent study (Li and Yang 2003) argued that a number of existing methods such as Rocchio, naive Bayes, and k-NN, can also be studied in this framework.

Although we present application case studies later in Chap. 8, we briefly mention two applications here, focusing on their use of predictive methods. Tan *et al.* (2000) discusses mining of call center records in an attempt to predict the cost of servicing a call. They explored an inductive learner (decision trees) and naive Bayes as classifiers. Weiss *et al.* (2000b) also discusses mining call center records, in this case to reduce a very large number of call center problem records to many fewer model cases.

A large number of developers are now working on filtering e-mail spam, and a number of companies sell products to filter spam from an e-mail stream. A conference on spam was held in Cambridge in January 2003. Graham (2003) discusses the use of Bayesian methods. Many other conferences on language processing have had papers on filtering spam, such as Kolcz and Alspector (2001), which discusses filtering using an SVM.

3.9 Questions and Exercises

1. Lets say you have a document collection that changes rapidly in a non-cumulative fashion and you wish to use it for classifying new documents. Which kind of classifier would be most appropriate for this task?
2. Can you think of what advantages the robust loss function has over the hinge loss function?
3. Give definitions of Precision and Recall for measuring the performance of a binary classifier.
4. Suppose you have a 11-NN binary classifier that has low recall score on a test sample of cases. Explain how the classifier could be modified to improve the recall score on this sample.
5. Try riktext on the train/test vectors created in Chap. 2 exercises for the category *earn*. Note that the category name specified on the command-line should be some string that does NOT occur in the dictionary. So you can't use "earn" with all lowercase characters. Use EARN or you can use E1 or CLASS or whatever. Note that you will need the dictionary used to create the vectors. Use the example in Sect. 3.2 of the riktext documentation as a guide. Use a default properties file and try a 2-fold cross-validation.
6. Use the program to examine other induced solutions. See if you can modify the rule file to manually improve performance on the test set. What is the problem with this manual approach to tuning rules?
7. Verify the probabilities in Fig. 3.17 and compute the class for a document that has words w1 and w2.
8. Try out the nbayes and testnbayes programs on the same train/test sets you used for riktext. Remember that nbayes and testnbayes are part of TMSK. Besides the vector files, note that you must use the same dictionary you used for riktext. Be sure you have the infile and vectorfile parameters set correctly in tmsk.properties. Look at the examples in Sects. 4.3 and 4.4 of the TMSK documentation to see what the command line should look like.

9. Now try the linear classifier on the train/test vectors. Use the programs linear and testline. These too are part of TMSK. Note that the files specified in tmsk.properties have to be in sync. For example, the index file (if specified) has to correspond to the vector file. Otherwise the program reads inconsistent data from different files and things blow up!! Note that the index file is an optional parameter. It can be safely commented out in the properties file. The main benefit of specifying the index file is somewhat faster execution. It is probably not noticeable in these assignments.

10. Try the linear classifier with tf*idf features. How do the results differ with the earlier solution?

11. Study precision-recall tradeoff in both linear and riktext. Discuss why precision goes down when recall goes up.

12. Generate a shorter dictionary from the current dictionary by taking the top 50 words. Generate vectors with this shorter dictionary and compare the performance of riktext and linear on the same train/test sets. Discuss the differences with the results from the larger dictionary.

Chapter 4
Information Retrieval and Text Mining

4.1 Is Information Retrieval a Form of Text Mining?

What is the principal computer specialty for processing documents and text? Many experts would respond "Information Retrieval." The task of information retrieval, or IR as its practitioners call it, is to retrieve relevant documents in response to a query. Figure 4.1 illustrates the objectives of information retrieval of documents, where (a) a general description is given of the query, (b) the document collection is searched, and (c) subsets of relevant documents are returned.

These seem like objectives far afield from predictive text mining. For prediction, the objectives are to (a) examine a collection of documents, (b) learn decision criteria for classification, and (c) apply these criteria to new documents. The goals of predictive text mining are illustrated in Fig. 4.2. These goals do not appear to match the goals of information retrieval.

The fundamental technique of information retrieval is measuring similarity. A query is examined and transformed into a vector of values to be compared with the measurements taken over the stored documents. In Chap. 3, we described prediction methods that use similarity measures for making decisions. The prediction problem is not solved directly by finding patterns in the collection of documents. Rather, similar documents are retrieved. We then look at these retrieved documents and only then measure their properties. Because we are interested in classification, we count the number of their class labels to see which label should be assigned to a new, unlabeled document.

We can now see that our objectives can be posed in the form of an information retrieval model, where documents are retrieved that are relevant to a query. Our query will be a new document. Like all documents, the query will be posed in terms of a word vector model. The query will be matched to all the stored documents, and a subset of documents will be retrieved. To make predictions, we add another step, as illustrated in Fig. 4.3. We must examine the properties of the retrieved documents, typically by simple criteria such as their labels.

In Chap. 3, we presented the nearest-neighbor methods, which are the embodiment of similarity-based prediction. We intentionally left out the details of measuring similarity and the potentially time-consuming search for similar documents.

S.M. Weiss et al., *Fundamentals of Predictive Text Mining,*
Texts in Computer Science 41,
DOI 10.1007/978-1-84996-226-1_4, © Springer-Verlag London Limited 2010

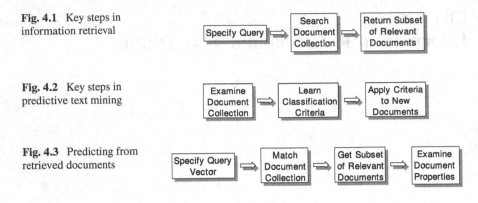

Fig. 4.1 Key steps in information retrieval

Fig. 4.2 Key steps in predictive text mining

Fig. 4.3 Predicting from retrieved documents

Because information retrieval has been studied as a separate endeavor, it would be wise to examine these "details" from both the IR perspective and the prediction perspective. This will allow us to make the best choices in applying a prediction method that measures document similarity. Although we have posed the prediction problem as a variant of information retrieval, there are differences and similarities among the full range of information retrieval techniques. We want to be aware of this relationship between information retrieval and prediction.

4.2 Key Word Search

Our technical goal for prediction is to classify new, unseen documents. We have made the case that information retrieval and prediction are unified by the computation of similarity of documents. Our most usual encounter with information retrieval is through a search engine, and our queries hardly match this paradigm of comparing documents by similarity. Yet, we should recognize that using a search engine is a special instance of the same general concept.

To invoke a search engine, we enter key words and expect relevant documents to be returned. These key words are words in a dictionary created from the document collection. In computing similarity among documents, almost all predictive techniques will use only the positive occurrences of words. While few, the words posed by the user of a search engine are the positive words of interest, so similarity can readily be computed.

The key words posed by a query to a search engine can be viewed as a small document. Therefore, it's just the usual document comparison, measuring how similar the new document (i.e., the query) is to the documents in the collection. Because the document is so small, our general expectations for comparing the similarity of documents are different from those for arbitrarily sized documents. Unless we have a very small collection of documents or uncommon key words are posed to the search engine, it is likely that an exact match of documents to the key words will be found. The notion of similarity is reduced to finding documents with the same key words as posed to the search engine. This creates special problems for a search engine because its ultimate objective is to rank the documents, not to assign a label. Thus, it may need additional techniques to break the expected ties, where all retrieved documents match the search criteria.

In the general prediction problem, the same techniques can be used to compute similarity. There, however, the number of words contained in the new document is expected to be much greater, and few if any exact matches will be found. Given just 100 words in a document, then 2^{100} possibilities might be hypothesized. Unless they are duplicates of highly related documents, they are likely to differ greatly. When comparing documents, we will not see exact matches but instead documents that are measured as being similar by having the highest scores.

4.3 Nearest-Neighbor Methods

From the perspective of statistical prediction theory, a method that compares vectors and measures similarity is a nearest-neighbor method. The heart of a nearest-neighbor method is its computation of distance between two examples, or in our case similarity. Because of the unique way of computing similarity and its specialized application to search engines, it is easy to see why the techniques are less attributed to classical prediction methods than to modern information retrieval theory.

The generalized methods are essentially the same for both prediction and information retrieval. However, their applications do diverge. For nearest-neighbor methods and prediction, the methods will collect the k most similar documents and then look at their labels. The actual documents are typically not of great interest and are not displayed in reaching a conclusion. The best value of k is usually determined by empirical experimentation. Although we may be tempted to accept the single most similar document, $k = 1$, there are strong theoretical reasons why this may lead to weaker results than using a larger number, especially when the sample is large.

For information retrieval, where classification is not the goal, the actual documents retrieved are of critical interest. They will determine whether a satisfactory response to the search query has been found. Moreover, we expect the best answers to be found near the top of the list of documents, where the similarity score is the highest, and it is more likely that an answer is relevant. Most users lose patience after reviewing the top ten answers, such as a summary of potential answers and their links found by a search engine. Most users will not go through tens of document links to search for an acceptable link. They instead change the query, thereby submitting a new document to compare to the collection.

The dimensions of the data will likely diverge. The document collection for a search engine is often much greater than needed for a binary classification problem, yet the number of words to be considered in computing similarity is far smaller. The query is a small document consisting of few words, and only these words are used in the computation of similarity. These are the positive words. For a typical prediction problem, the comparison is of a complete document, where thousands of words can be positive.

Still, these methods are conceptually the same. They all compute similarity measures. Let's look at how documents are compared.

4.4 Measuring Similarity

Many measures of similarity could be specified. We will consider three increasingly complex measures. These are intuitively clear and are a natural progression from simple to complex along the same theme. Although these measures are not shown to be always best, they have demonstrated their value through many applications. Given two documents, we will examine how similar they are. The result of these efforts is a numerical measure of similarity. In general, we compare one document to a collection of documents. Therefore, we have a list or vector of similarity measures, one for each document. If we are interested in just the highest k measured documents, then we might just keep a running list of k documents and their similar measures. Let's see how to compute similarity between documents.

4.4.1 Shared Word Count

The most obvious measure of similarity between documents is a count of their shared words. Which words are we talking about? For an information retrieval system, we likely have a global dictionary, where all potential words will be included in the dictionary, with the exception of stopwords. For prediction, it is generally better to preselect the dictionary relative to the label. We typically think of computing similarity independently of the labels and only afterwards examining the labels of the retrieved documents. Yet, the labels can be valuable in composing a local dictionary, such as the topic of some newswires.

Figure 3.10 from Chap. 3 is an illustration of the process of computing similarity by shared words. We look at all the words in the new document and for each document in the collection count how many of these words also appear. No weightings are used, just a simple count. If the initial dictionary has true key words, with weakly predictive words removed, then this count will produce reasonably predictive results. The great value of this measure is that the results are clearly intuitive. No one will question why a document was retrieved or how the measure was computed.

The vectors for each document can be described in terms of zeros and ones. Mathematically, the similarity of two documents is the product of the two vectors because only when two ones are multiplied is a word counted. In terms of implementation, one can readily match ones by AND-ed bit manipulation operations. Any way you look at it, computing shared words is easy. The quality can be good for a well-composed dictionary, but performance can degrade with larger dictionaries containing nonspecific words. We next look at a minor modification that sometimes substantially increases predictive performance.

4.4.2 Word Count and Bonus

In high dimensions, it's difficult for a nearest-neighbor method to readily discriminate between predictive and weakly predictive words. Although we may give the

Fig. 4.4 Computing
similarity scores with bonus

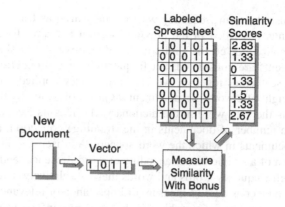

method some help by carefully selecting the dictionary words in advance, it is advantageous to use more than the presence of words in the similarity measure. Instead of computing similarity based on binary features, we might compute similarity as suggested in (4.1) and (4.2) for computing the similarity between a new document that contains K words and document $D(i)$.

$$\text{Similarity}(D(i)) = \sum_{j=1}^{K} w(j), \tag{4.1}$$

$$w(j) = \begin{cases} 1 + 1/\text{df}(j), & \text{if word } (j) \text{ occurs in both documents,} \\ 0, & \text{otherwise.} \end{cases} \tag{4.2}$$

We compute a measure of similarity between two documents for all dictionary words occurring in both documents. As before, a value of 1 is added to the similarity measure when a word is positive in both documents. A bonus is also added to the measure. The bonus is $1/\text{df}(j)$, where $\text{df}(j)$ is the number of documents in which the word j occurs in the collection, a variant of idf, the inverse document frequency. Thus, if the word occurs in many documents, the bonus is small. If the word occurs in few documents, the bonus is relatively large. Instead of a 0 or 1 contribution for each word, we now have a 0 to 2 contribution, where the bonus also varies between 0 and 1. Figure 4.4 is an example of this similarity computation.

We have used this simple measure in at least one significant predictive text-mining application. It does significantly better than the simple shared word count. It does help discriminate among the weak and strong predictive words, yet maintains an intuitive feel. On that same benchmark, the widely used cosine distance did better by a couple of percentage points. We look at cosine distance next.

4.4.3 Cosine Similarity

The classical information retrieval approach to comparing documents is cosine similarity. The word count and bonus approach is a variation of computing tf-idf. Because the term frequency was just measured as true or false, the word count and bonus described in the previous section are already normalized and scaled reasonably. For cosine similarity, only positive words shared by the compared documents are considered, but frequency of word occurrence is also valued. Putting these themes together, we have the widely applied cosine distance of (4.3). The weight of a word in a document $w(j)$ is computed by the tf-idf formulation, where j is the j-th word in the dictionary, tf(j), is its frequency in the document, N is the number of documents in the (training) collection, and df(j) is the number of documents in which the word appears. As discussed in Chap. 2, tf-idf is an extension of a simple binary encoding of word presence, and extra weighting is given to high-frequency words and words that are relatively unique. The tf-idf measure can also be considered a measure of importance or relevance to the document. Because documents are of variable length, frequency information could be misleading. The tf-idf measure can be normalized to a unit length of a document D as described by norm(D) in (4.3). The value of norm(D) is a constant; it can be computed and stored in advance for each document. Note that $w(j)$ is always zero when a word does not appear in a document, so that only shared words between two compared documents are of interest. The cosine distance multiplies the weights of the shared words of the two compared documents, which is similar to a logical AND operation requiring a word to be present in both documents. The measures of individual words are summed over all words according to (4.3), resulting in a measure of overall similarity between two documents.

$$w(j) = \text{tf}(j) * \log_2(N/\text{df}(j)),$$

$$\text{norm}(D) = \sqrt{\sum w(j)^2}, \tag{4.3}$$

$$\text{cosine}(d1, d2) = \sum (w_{d1}(j) * w_{d2}(j))/(\text{norm}(d1) * \text{norm}(d2)).$$

The multiplications to compute similarity ensure that only positive words in both documents will be addressed. This also leads to efficient computation with inverted lists, which will be described in Sect. 4.7. By including frequency information and normalizing the computation, the similarity measures may be less than fully intuitive and wander over a wide range, but the results have proven their worth over time. Cosine is the default computation for information retrieval and should serve as a benchmark for improvement in any application. Nearest-neighbor methods for prediction do not assume any fixed method of computing distance or similarity, and results may be improved by trying alternatives and subjecting them to rigorous evaluation.

4.5 Web-based Document Search

These measures for computing similarity are quite satisfactory for classical information retrieval. For searching the Web, which is our most common encounter with their application, they are only partially fulfilling in finding the most relevant documents.

In comparing prediction and information retrieval, we have described several differences in the applicability of similarity measures. For prediction, we expect the query to be a document with many words. For information retrieval, the query is typically a few key words. Does this make a difference? For prediction, the goal is to classify, to find the correct label. If the key words are good, even if they match many documents, the process can be successful because we only need to count the labels of the retrieved documents. For information retrieval, our objective is to order and examine the retrieved documents. The Web is composed of a massive number of documents. Unless the key words are unique, a very large number of documents may be retrieved having all key words. Because the similarity scores are likely to be nearly identical, there is no way to order the retrieved documents based on these similarity measures.

Until search engines like Google arrived, it was not unusual for a search engine to present thousands of potential answers to a query, but the ordering of these retrieved documents was most unappealing. It would be the responsibility of a user to refocus the query by adding more words or by substituting alternative, more specific words. Although the document space of the Web is massive, these documents contain more information than just words. Web documents are linked to other documents, and this information can be exploited to rank the documents retrieved in response to a query. An effective ranking of documents will improve predictive performance, providing a more accurate group for the k-nearest neighbors.

Formally, a page-ranking function assigns a score to each Web page based on the query. The larger the score, the more relevant the page is to the query. Mathematically, the page-ranking problem can be posed as a prediction problem. For example, given a query, we may predict the degree (or the probability) of each Web page's relevance to the query. Therefore, given enough training data, one may in principle use machine-learning methods to learn a page-ranking function. An important issue in Web searching is the difference between query-independent ranking and query-dependent ranking of Web pages. Ideally, the ranking score of a Web page is the relevancy of the Web page to the query, and thus should be query-dependent. However, some algorithms assign a score to each Web page independently of the query. In this case, the score can be interpreted as an indication of the page's quality. After a search engine retrieves a large number of potential answers, the highest-quality pages can then be presented first, followed by the lower-quality pages.

Information about links will help in ranking, which is needed to select the best k documents within a larger set of retrieved documents. How then can we rank effectively when only a few key words are in a query?

4.5.1 Link Analysis

For a simple query using any of our distance measures, we will get lots of ties whether we use the direct match of the query words or a more sophisticated cosine

Fig. 4.5 Computing the
PageRank

> 1. Use (4.4) to compute PageRank for each page in the
>   collection using latest PageRanks of pages.
> 2. Repeat step 1 until no significant change to any
>   PageRank.

distance. Unless the words are unique, all distance measures will return large numbers of responses. One of the insights of modern search engines such as Google has
been to use additional information about citations to rank these tied responses.

To solve this problem, we hold an election, where each document in the collection votes for its favorite documents. For the Web, the documents are Web pages.
This concept of voting is not that unusual. Scientific papers cite other papers that
they deem relevant. The importance of a collection of papers can be ranked objectively by looking at these papers and measuring how often a paper is cited by other
papers, most significantly by other highly ranked papers. Similarly, Web pages have
links to other Web pages, and a similar citation analysis can rank the Web pages
by importance. The overall concept is simple. When a large group of Web pages
are retrieved, ordering is needed. Most people will examine a small number of documents before dismissing the list as unsatisfactory. The documents that should be
ranked highly are those that are voted most relevant by other documents. These are
the most likely to satisfy the user who has posed a somewhat nonspecific query.

How are these rankings computed? Let's look at the PageRank algorithm of the
Google search engine. The computation is surprisingly simple even over a huge
number of pages. A numerical measure or rank will be assigned to each page on
the Web, corresponding to a linked document. Let A be a page and T_j be a page
pointing to A. The page rank of page A, $PR(A)$, is given in (4.4), where d is a
minimum value assigned to any page and $C(T)$ is the number of outgoing links
from page T. Thus we see that the rank of any page is dependent on the number
of links pointing to it and their ranks. It's a recursive definition. A page will be
ranked highly when many highly ranked pages point to it. To compute the ranks, the
trivial iterative algorithm of Fig. 4.5 is employed. We proceed sequentially through
an ordered list of pages, repeatedly computing the page rank of each page using the
most recent computed values of its linked pages. If all pages with no outgoing links
are discarded, the average PageRank will be 1.

$$PR(A) = d + (1 - d) * \sum_j (PR(T_j)/C(T_j)). \qquad (4.4)$$

The algorithm can be illustrated using the example in Fig. 4.6, where Page A can
be seen as a home page that points to pages B and C. Page B and Page C each point
back to A. Therefore A has two links leaving and two links entering, and B and C
have one link entering and one link leaving.

Table 4.1 shows the progress of the iterative PageRank computations. We arbitrarily assign initial values of 1 to pages A, B, and C. Using a value of $d = .1$, in
the first iteration, we compute as follows:

$$PR(A) = .1 + .9 * (PR(B) + PR(C)) = .1 + .9 * (1 + 1) = 1.9,$$

Fig. 4.6 Example of links
between pages

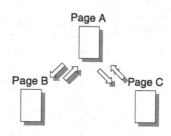

$$PR(B) = .1 + .9 * (PR(A)/2) = .1 + .9 * (1.9/2) = .95,$$

$$PR(C) = .1 + .9 * (PR(A)/2) = .1 + .9 * (1.9/2) = .95.$$

The process continues through 25 iterations, where the answer stabilizes at $PR(A) = 1.48$, $PR(B) = .76$, and $PR(C) = .76$.

The PageRank algorithm assigns a score to each web-page that can be interpreted as the quality of the web-page. Variations of page-rank have also been employed in web-search. For example, if we apply this algorithm at the host level instead of page level, then we can obtain a quality score for each host which can be interpreted as the trust-worthiness of the host. Another variation is to apply PageRank to a subset of web-pages focusing on a certain topic, and the score of a web-page indicates the quality of this web-page specialized to the topic. In particular, one can apply link analysis to the collection of pages obtained from search results, which are pages related to the search. This leads to search specific page quality scores. A different algorithm, called *Hubs and Authorities algorithm*, can also be applied to search results. In this algorithm, each page has two scores: a hub-score and an authority-score. Authorities are web-pages that are authoritative on a special topic, and Hubs are web-pages that link to many authoritative web-pages on a special topic. A web-page has a high authority score if it is linked by web-pages with high hub-scores; a web-page has a high hub-score if it links to web-pages with high authority-scores. Using this idea, we can iteratively update the authority-score and the hub-score of every web-page: update the hub-score of each web-page as the sum of authority-scores (of web-pages) linking to it; update the hub-score of each web-page as the sum of authority-scores (of web-pages) it links to. This iterative procedure is rather similar to that of PageRank.

Another popular method of using link structure is based on the observation that the anchor text of a link describes its target. Anchor text is the "clickable text" associated with a hyperlink that is highlighted in an HTML (hypertext markup language) page. Most current search engines associate the anchor text of a link with the page it points to. One may also include ALT text for image hyperlinks when such text is available. As an example, in a hyperlink <A HREF="http://www.whitehouse.gov"> White House</A>, the anchor text is "White House." We associate "White House" with the URL "http://www.whitehouse.gov." Since anchor text is typically created by humans for the purpose of describing the linked target, it often gives a high-quality summary of the page it points to. The effect of anchor text is quite similar to the title of a document in that a very small number of terms are used to capture what

Table 4.1 Iterative computations of PageRanks

Iteration	PageRank(A)	PageRank(B)	PageRank(C)
0	1.00	1.00	1.00
1	1.90	.95	.95
2	1.82	.92	.92
3	1.75	.89	.89
4	1.70	.87	.87
5	1.66	.85	.85
6	1.62	.83	.83
7	1.59	.82	.82
8	1.57	.81	.81
9	1.55	.80	.80
10	1.54	.79	.79
11	1.53	.79	.79
12	1.52	.78	.78
13	1.51	.78	.78
14	1.50	.78	.78
15	1.50	.77	.77
16	1.49	.77	.77
17	1.49	.77	.77
18	1.49	.77	.77
19	1.48	.77	.77
20	1.48	.77	.77
21	1.48	.77	.77
22	1.48	.77	.77
23	1.48	.77	.77
24	1.48	.76	.76
25	1.48	.76	.76

a document is about. However, an advantage of using anchor text is that there can be multiple people generating different anchor texts for a particular page. Therefore, relevant query words missing in the title may still appear in one of the anchor texts.

The popularity and success of search engines that use link analysis demonstrate the importance of looking beyond word analysis for some information retrieval tasks. Link analysis is an evolving and often proprietary technology. Based on experience, designers of search engines include many special hooks to achieve their goals. For competitive reasons, not all these special tweaks are revealed. Link analysis is important, but other variations have evolved. For example, pages that are connected by links are more likely to have the same topic than those that are not. This observation can be used to rank or organize search results. Alternatively, documents may first be categorized, and votes from the same category are given a much higher weight. The main message is clear. If we can compile a measure of popular-

Fig. 4.7 Generalizing search to document matching

ity or frequency of citation of documents, we can clearly improve our performance in document retrieval.

4.6 Document Matching

We have contrasted search engines and prediction for text. One difference is the size of the documents, and another is the goal of assigning a label or examining the retrieved comments. These differences are not firm. A generalization of searching is document matching. As we described for prediction, a complete document is matched, not just a few key words. For document matching via a generalized search engine, an arbitrarily long document is the query, but the goal is still to rank and output an ordered list of relevant documents. The most similar documents are found using the measures described earlier. Figure 4.7 shows the generalization of a search engine to document matching.

Why is it interesting to match complete documents? When only a few words are presented, a document matcher behaves like a search engine, where a user can add words, thereby specializing the search until finding a satisfactory set of responses. But matching a long document can also be interesting. Consider an online help desk, where a complete description of a problem is submitted. That document could be matched to stored documents, hopefully finding descriptions of similar problems and solutions without having the user experiment with numerous key word searches.

4.7 Inverted Lists

Matching a document to a collection of documents looks like a tedious and expensive operation. Even for a short query, comparison to all large documents in the collection implies a relatively intensive computation task. Information retrieval techniques do function in reasonable time and have been developed to mitigate the complexity of sequential comparison of all documents in a collection. The query's words are sparse relative to the dictionary's words. Of greatest benefit is the use of an inverted list. Instead of documents pointing to words, a list of words pointing to documents is the primary internal representation for processing queries and matching documents.

Figures 4.8 and 4.9 illustrate the concept of focusing on words rather than documents. The natural representation for predictive modeling was described in Chap. 2.

Fig. 4.8 Documents pointing
to words

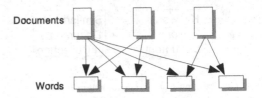

Fig. 4.9 Words pointing to
documents

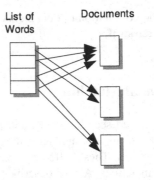

Fig. 4.10 Document order
and word order

Document Order		Word Order	
Documents	Words	Words	Documents
1	5, 100	...	...
2	100, 200	5	1, ...
...	...	...	...
		100	1,2, ...
		...	...
		200	2, ...
		...	...

Each document in the collection is represented as a list of its constituent words. Nearest-neighbor methods compare a new example to stored examples. For information retrieval, our representation is more specialized. A general distance measure is not computed over all attributes. Instead, the measures of similarity for IR are all based on positive co-occurrences of query and matched document words. So if a word is absent from the query, it will have no contribution to the similarity measure. Moreover, the only documents of interest are those where a query word appears.

It is trivial to convert from a document order to a word order. In Fig. 4.10, an example with both orders is shown. The conversion from one order to the other is straightforward.

When a query is presented, we process each of its words. The word list tells us the documents where the word appears. As the word list is processed, the similarity measure for the indicated document is increased by the requisite amount. If we were using just a binary count, then for each word on the list, a 1 would be added to the similarity measure of that document. In our example, if the query contained words 100 and 200, then we would proceed through the following steps, first processing

$W(100)$ and then $W(200)$ to compute the similarity $S(i)$ of each document i:

$$S(1) = 0 + 1,$$
$$S(2) = 0 + 1,$$
$$\ldots \text{Other documents with } W(100) \ldots$$
$$S(2) = 1 + 1,$$
$$\ldots \text{Other documents with } W(200) \ldots.$$

The inverted list is the key to the efficiency of information retrieval systems. Similarly, for purely predictive methods, the same efficiencies can be adopted and implemented. We see that search engines operate quite efficiently. Other methods such as the linear score are more efficient for predictive applications because they have trained and found a mathematical formula to apply. The expectation of sparseness, the positive-match similarity measures, and the inverted list have all contributed to making nearest-neighbor methods a pragmatic possibility for prediction. An implementation of document matching is available in the accompanying software.

4.8 Evaluation of Performance

Information retrieval methods are specialized nearest-neighbor methods, which are well-known prediction methods. IR methods typically process unlabeled data and order and display the retrieved documents. For prediction of a new document, a supplementary step augments the usual retrieval. The labels of the k most similar documents are counted, and a class assignment is made. In the context of labeled prediction, the methods described in Chap. 3 for evaluation are applicable. For example, a test set of documents can be used to measure an error rate.

For most applications, the date of appearance of documents is important. The most appropriate testing is to look at documents appearing at an earlier date and test on a later date. For example. if we were trying to determine whether the stock market would go up or down, the documents that we examine should be from a period prior to when the stock price is known. There may be applications where time is less important. For these applications, information retrieval methods would have a natural way of evaluating predictive performance. Unlike other learning methods such as linear scoring, the IR methods have no training and induce no new rules for classification other than determining the value of k, the number of nearest neighbors to be used in counting the labels. For each document from the collection, or from a random subset, the nearest neighbors could be retrieved along with their labeled answers. The accuracy for a specific value of k could be determined.

Both measuring accuracy and selecting k during the same experiments are slightly optimistically biased even when the document collection is large. It's the most pragmatic approach for a real application. Finding k over many classes is a much more tedious effort. For example, to assign topics, such as finance or sports, to newswires might require finding a different k for each topic. Such an approach is not very practical and is questionable when the same test data are reused. Instead, the same value of k is usually selected for most classes, perhaps with some allowances made for the rare classes. Couldn't we just use the label of the single most similar document? In general, that answer will be weaker than using more than one document.

Search engines and document matchers are not focused on classification of new documents. Their primary goal is to retrieve the most relevant documents from a collection of stored documents. How do we measure whether that goal is achieved? We could ask users to rate the results of a search engine and we would apply some measure of accuracy. For example, we may ask how many of the top k (typically top 10) pages returned are relevant. This effort is a form of assigning labels. For the Web, the user community measures effectiveness by choosing from among the many search engines. Newer search engines have displaced the originally dominant ones because users have switched *en masse* to the search engines that gave them the best results. The evaluation is by voting of a user community. Interestingly, the search engines that win in the marketplace simulate this voting experience by using link analysis as a proxy for the vote of the user community.

4.9 Summary

Information retrieval is described in terms of predictive text mining. The methods can be considered variations of similarity-based nearest-neighbor methods. Both key word search and full document matching are examined. Different methods of measuring similarity are considered including cosine similarity. Classical information retrieval has evolved from retrieval of documents stored in databases to web or intranet based documents. These document have richer representations with links among documents. Link analysis for ranking similarity of documents is described. Some performance issues for computing similarity are considered including the specification of inverted lists for indexing documents.

4.10 Historical and Bibliographical Remarks

Many of the ideas discussed above have been part of the technical repertoire of information retrieval specialists for many years. The SMART system (Salton 1964; Salton and Lesk 1965, 1968) incorporated the vector representation of documents and queries and the idea of using the cosine of the angle between vectors as a correlation coefficient. The system also had the ability to rank answers according to the value of the correlation coefficient. Evaluation of the various ways to configure the SMART system was measured by precision and recall and the 10-point precision–recall curve (Salton and Lesk 1968). The idea of weighting query terms by a combination of term frequency and inverse document frequency is explained in Salton and Wu (1980).

The PageRank algorithm, discussed in Page and Brin (1998) and Page *et al.* (1998), is closely related to the *Hubs and Authorities algorithm* (Kleinberg 1999). This method for ranking pages for output is the essential idea in the Google retrieval engine (but is not the complete story, which is proprietary). The basis for this algorithm is another old idea in information retrieval called "citation analysis," pursued for many years by Garfield (1972).

Research in Web-based information retrieval has been supported by the TREC (Text Retrieval Conference) program in recent years. The detailed descriptions of the program, with proceedings of the previous years, can be found at http://trec. nist.gov. There are two main Web-retrieval-related evaluations, one for *home-page finding* and the other for *topic distillation*. The goal of home-page finding is to return the main page of a site. For this task, link analysis is quite useful. The goal of topic distillation is to find relevant pages for more specific query topics. In this task, a query is addressed by the content of relevant pages. Therefore, traditional IR-based retrieval methods work quite well, and link analysis seems less useful.

In addition to the simple document-matching method we described in this chapter, more sophisticated methods have also been proposed. They can be regarded as ranking functions that determine the relevance of a document to a query. For example, one such formula, called BM25 (Robertson *et al.* 1994), has been quite successful in some TREC evaluations. However, a disadvantage of such methods is that they often contain a number of parameters that need to be tuned. In principle, one may also use machine-learning methods to learn unknown parameters of a ranking function or to combine the outputs of different ranking functions. Given a query, we need to predict whether a document is relevant to a query or not. The ranking score of a document can then be interpreted as the degree or probability of its relevance to the query.

4.11 Questions and Exercises

1. Verify the results of Table 4.1. You can write a small program based on Fig. 4.5 or use an EXCEL spreadsheet to do this.
2. Take any business article from your favorite online newspaper, and create an XML file by tagging it the same way as the Reuters articles. You can do this as simply as cut/paste of article text and adding the relevant tags. You can unpack the Reuters corpus into a temporary directory and see the individual articles to see examples of what you must create. Now use this file to retrieve the 10 best matches from the Reuters training XML documents using the TMSK routine *matcher*. Comment on the quality of the matched results. Note that you must have created the inverted index file for the training vectors—see the TMSK documentation. What are the key determining factors for the quality of the matched results?
3. Why does the cosine distance need to be normalized for document length?

4. Discuss how the page-rank (4.4) is related to the behaviour of an Internet surfer.
5. What do you suppose is the most common anchor text? Find at least one other instance of misleading anchor text on the web.
6. Assume that a human assessor is available to evaluate results of a matcher. Compare and contrast two different methods for validating a matcher.

Chapter 5
Finding Structure in a Document Collection

Prediction methods look at stored examples with correct answers and project answers for new examples. One would expect that if we cannot obtain answers for the training examples, then the process cannot be completed. Given a collection of documents, we have no problem transforming the unstructured set of words for each document into a structured spreadsheet. But the last column also must be filled in. In Fig. 5.1, we see a spreadsheet, a list of labels, and the spreadsheet column containing the labeled answers. Someone must compose a list of potential labels. Given the list, someone assigns labels to the documents. Sometimes label assignment can be automated, such as the label that a company's stock price has risen. In most instances, such as topic assignment to newswire articles, the assignment of labels is done by humans, and this can be a tedious and expensive task. Is there any way to assign labels automatically to a document collection? We will discuss this task. Not only will the labels be assigned, but the list of labels will also be determined automatically. Because such key information is missing from the problem description, our expectations for accurate predictive performance should be reduced from standard prediction applications with labeled data.

Our objectives are to determine labels and assign them to documents. If we were very confident of meeting these goals by some automatic process, we would completely bypass the expense of having humans assign labels, but the process known as document clustering is less than perfect. The labels and their assignment may not be the same as those composed by humans or those collected from some objective process that uses external information such as stock price change. Document clustering assigns each of the documents in a collection to one or more smaller groups called clusters. Based on an examination of their words, these clusters should contain similar documents. The initial collection is a single cluster. After processing, the documents are distributed among a number of clusters, where ideally each document is very similar to the other documents in its cluster and much less similar to documents in other clusters. Figure 5.2 illustrates the overall task of taking a complete collection and assigning its documents to smaller clusters.

It is quite reasonable to expect that documents having the same label will be similar. If a human composed the labels, these goals were likely known prior to

S.M. Weiss et al., *Fundamentals of Predictive Text Mining*,
Texts in Computer Science 41,
DOI 10.1007/978-1-84996-226-1_5, © Springer-Verlag London Limited 2010

Fig. 5.1 Spreadsheet with labels

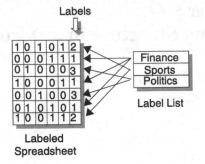

Labeled
Spreadsheet

Fig. 5.2 Clustering a document collection

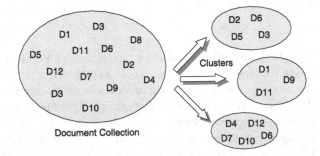

Document Collection

assembling the structured data. The labels have an important meaning, often identifying the goals of the project. Given a list of labels, the examples represented in
the sample were then assembled and the labels assigned. Prior to document clustering, the labels are unknown. Instead, a measure of similarity is used to group the
documents. Only after clustering are the groups examined to give meaning to each
cluster. The clustered documents do form groups, but their meaning is implicit in
the grouping. The process did not start with a good understanding of specific goals.
So only after it is completed and the clusters have been reviewed can we say that
something useful has been achieved.

For an application, there are many ways to organize data into groups. Consider
the following experiment: Take a collection of labeled documents, and then cluster them while ignoring the labels. Will the document clusters number the same as
the originally labeled classes? Will each cluster be pure so that all of its documents
have the single original label? In general, the results will not match exactly. In many
applications, there will be some clusters that are very close to the original classes
but others that are far away. This doesn't mean that the clustering process is unsuc-
cessful. Its objective is to group the documents by similarity. Its notion of similarity
may not be the same as, and may not meet the objectives of, the composer of the
original labels.

If document clustering is so murky, why would anyone attempt it? Clustering
gives additional structure to the sample data, and that can be useful. In the best case,
the clusters relate to a goal that is similar to one that would be attempted with the extra effort of manual label assignment. In that case, the label is an answer to a useful

question. If we are at a stage where the question has to be formulated, then the process of document clustering can be informative. For example, suppose a company is operating a call center where users of their products submit problems, hoping to get a resolution of their difficulties. The queries are problem statements submitted as text. Surely the company would like to know about the types of problems that are being submitted. Clustering can help us understand the types of problems submitted. Printer problems might cluster in one group and network problems in another. For a fixed number of clusters, such as 100 different clusters, someone could review each of the clusters to find the most frequent types of problems received by the call center. These might be characterized by identifying key words found for a cluster.

5.1 Clustering Documents by Similarity

If our goal is prediction, then similarity measures, as embodied in information retrieval and nearest-neighbor methods, are usually not the first choice. They are unable to simplify a solution by directly learning from labeled examples. Yet, when we discuss document clustering, similarity measures assume a prominent role. Here the data are unlabeled and similarity of documents is the defining characteristic for assigning labels. So, to cluster documents, it is natural to reexamine the information retrieval techniques and their integral similarity measures.

To cluster documents, it will be necessary to compare documents and group together those that are similar. As shown in Fig. 5.3 and as is typical of information retrieval, two documents are compared and a measure of similarity is computed. For document clustering, the scenario in Fig. 5.4 is more common. A composite document is compared with another composite document. The composite document represents many documents either in summary form or by their individual member documents. How do we represent a composite document? Some clustering methods, such as hierarchical clustering, maintain links among the composite document, the cluster, and all their member documents. Other methods, such as k-means clustering, summarize the composite cluster by classical statistics using averages of individual measurements.

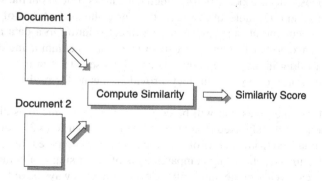

Fig. 5.3 Computing similarity of two documents

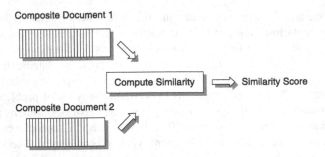

Fig. 5.4 Computing similarity for clustering

Each of these approaches has advantages and weaknesses. Relative to prediction, we are operating in a fog, so that predictive accuracy is not necessarily the deciding factor in choosing an approach. There is much research literature on clustering documents. For our purposes, a few prominent methods are sufficient to highlight contrasting approaches to comparing single or composite documents.

5.2 Similarity of Composite Documents

Information retrieval methods can dynamically obtain the documents most similar to any other document. So, for many applications, why bother to predefine groups of documents? These relationships may be found dynamically by matching a document with others and measuring their similarity. For example, some attempts have been made to add a clustering component to responses of a search engine. The idea is to group the answers, the linked documents, into different categories so that the user can get an overview of all answers instead of processing sequentially through the first 10 or 20 documents. In general, though, the supplementary clustering information has had limited success because the ordered response of a search engine may trump a clustered view of all documents, including the weaker ones.

We have mentioned some applications where it is beneficial to organize the documents by groups. This is especially true when one knows little about the structure of the documents and one wants an overview of their contents. In terms of prediction applications, document clustering provides a direct solution to a data acquisition problem. We have no labels. Our spreadsheet is missing a column, and this is a rational way of filling in the last column. Although these entries in the column may be answers in search of a problem, they provide a means for further analysis and prediction.

The similarity measures that will be considered are the same ones discussed in Chap. 4. The most widely used is cosine similarity, shown in (4.3) from Chap. 4, where the normalized tf-idf vectors of two documents are compared. Instead of comparing two documents, clustering compares pairs of either single or composite documents. How can we compute similarity? Several obvious ways have been used to represent comparisons among composite documents.

The simplest way to summarize a cluster is to create a composite document by averaging the documents in the cluster. Each document is represented by a vector of measurements. These might be the normalized tf-idf measurements. For each measurement in the vector, the mean can be taken. The result is a single summarizing vector for a cluster. Thus we have a composite document represented in the usual vector format. Comparing another document with the mean vector of the composite document is the same process as comparing two documents. In text analysis, one often uses a normalized document vector representation. Similarly, one can normalize the mean vector of the composite document. For cosine similarity, the two vectors are multiplied, assuming they are normalized.

The cluster was summarized by averaging the vectors of all documents in the cluster. This results in a new vector. It can be compared with any single document or with other clusters that also have mean vectors. Alternatives can be considered for the summarizing vector. Each of the constituent documents that are members of the cluster can be examined, and they could be candidates for a summarizing vector to characterize the cluster. Given a document not in the cluster, our objective is to find its similarity to the documents in the cluster. We can compare the new document to each of the documents in the cluster and note their similarity. We can also compare two clusters by comparing all pairs of documents, one from each cluster. Then we measure similarity by the similarity of the best pairs. Which are the best pairs? Two measures have been used:

- single link: the two most similar documents,
- complete link: the two least similar documents.

We need not make any distinction between single documents and composite documents. A single document could be considered a cluster of one document. For a single link, when we compare all clusters, the global best pair is the two most similar documents from any two different clusters. For a complete link, similarity is measured in two steps. First, we select a pair of clusters to compare. For each pair of clusters, the measure of similarity is the two least similar documents, one from each cluster. The result of the first step is a single similarity measure for each pairwise comparison of clusters. In the next step, we find the single best pair, which is the one with the greatest similarity. The two most similar clusters are the two clusters with the greatest minimum pairwise similarity.

Measures of similarity for composite documents are illustrated in Fig. 5.5, where the first composite vector has three vectors and the other has one. In this example,

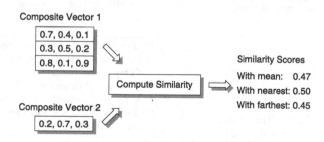

Fig. 5.5 Example of similarity computation for composite documents

similarity of two vectors is computed by their inner product, i.e., multiplying the non-normalized components of each vector and summing over all components. The mean vector is a one-time computation for the cluster, and the resultant new vector is immediately available for direct comparison with other vectors. Using this averaged vector is considered a faster comparison than comparing individual pairs from different clusters.

We will look at one method, k-means, that requires using the mean. For hierarchical clustering, any of the three measures can be used. Knowledge of the application and efficiency will strongly influence the choice of clustering method.

5.2.1 k-Means Clustering

k-means is a classical clustering method that has been adapted to documents. It is very widely used for document clustering and is relatively efficient. The concept is illustrated in Fig. 5.6, where the documents start in one pile and are then distributed into smaller piles of similar documents. For purposes of prediction, each pile of documents can be considered a unique label.

How do the documents end up in the right cluster? The name k-means implies that k clusters will be used, and the means of these clusters will play an important role. Figure 5.7 describes the algorithm. Documents are moved among a fixed number of bins until no improvement can be found. How are the documents initially distributed among the bins? The classical approach of k-means is to randomly assign the documents to the bin. Because the algorithm halts at a local minimum, several runs with random assignment may be necessary to get the best answer. An alternative approach that we have used is to compute the global mean vector of all documents, compute the similarity of each document to this vector, sort the similarity measures and corresponding documents, and assign equal numbers of documents

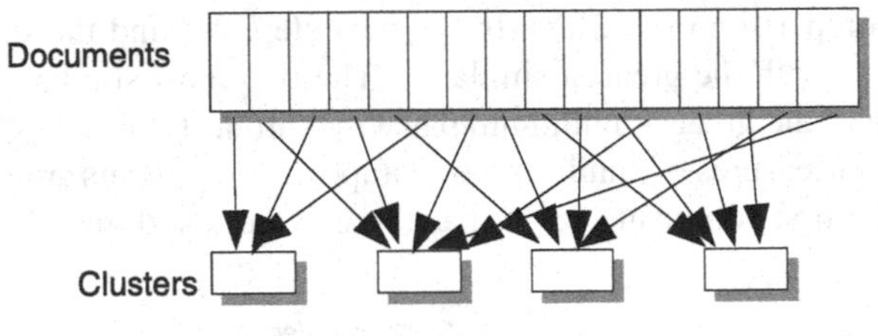

Documents

Clusters

Fig. 5.6 Assigning documents to clusters

Fig. 5.7 The k-means clustering algorithm

1. Distribute all documents among the k bins.
2. Compute the mean vector for each bin.
3. Compare the vector of each document to the bin means and note the mean vector that is most similar.
4. Move all documents to their most similar bins.
5. If no document has been moved to a new bin, then stop; else go to step 2.

Fig. 5.8 Example of
k-means clustering (for
k = 2)

Documents in Clusters

	Cluster 1	Cluster 2
Initial:	0,4,2,3,4	

	Cluster 1	Cluster 2
Step 1:	0,4	2,3,4
	mean=2	mean=3
Step 2:	0,2	4,3,4
	mean=1	mean=3.67
Step 3:	0,2	4,3,4
	mean=1	mean=3.67

in sorted order to the bins. The k-means algorithm is then applied, but only moves between the adjacent bins are considered for moving a document.

Note that if the original documents are represented as normalized vectors, then, in step 2, the mean vector can also be normalized. Figure 5.8 is an example of the application of k-means for two clusters, five documents, and hypothetical vectors containing only a single word measured by frequency.

k-means will typically converge to its minimum after relatively few iterations. When the number of bins is much smaller than the number of documents, which is the expected case, the algorithm can be considered efficient. As the number of bins grows large, efficiency diminishes. At the limit each bin is a single document and every document is compared with every other document during each iteration.

The main conundrum for k-means clustering is to determine k, the number of clusters. The method has no knowledge of the right k, so it is our burden to specify k. How do we do this? Very often, for an application, we have a good idea on a relative basis for choosing k. We may know that five or ten bins are all that are reasonable. If we expect great diversity, such as for our call center example, we might choose a larger number like 100. If we know something about the nature of the documents, we may also know something about a reasonable number of labels or categories. In general, the number of bins should be far smaller than the number of documents. Otherwise, prediction and generalization to new documents will be weakened.

Is determining k strictly guesswork? For k-means, we can use some of the empirical techniques that were described in Chap. 3. The k in k-means is similar in concept to the k in k-nearest neighbor. It is a complexity measure that we can estimate from the sample. For the k-nearest neighbor method, we choose k based on the value that minimizes test error. For clustering, an alternative measurement, an objective function, is employed to measure overall performance. For k-means, an objective function analogous to error rates can be described. Let x^i be the i-th document vector, and $c_i \in \{1, \ldots, k\}$ be its corresponding cluster index, a typical measure is the total variance from the cluster means as described in (5.1) for k clusters.

$$E(k) = \sum_{i=1}^{n} \frac{(x^i - m_{c_i})^2}{n}. \tag{5.1}$$

For each document vector x^i and its cluster mean m_{c_i}, compute the squared error averages over all n documents, $E(k)$. Then we vary k, perhaps by starting with two clusters, then four, and continuing to rerun k-means and doubling the number of clusters, stopping when k is too large. We then compare the objective function values for varying k. We choose k based on the point where increasing k does not yield a commensurate decrease in variance.

After the algorithm is completed, a set of labels can be assigned to the original spreadsheet. At that point, we have a standard classification and prediction problem. That implies that any learning method for classification could be applied. We should not forget the pedigree of this method. Because the labels were assembled by looking at the mean vectors of cluster documents, the natural prediction method is to use the same technique to classify and predict new documents. If we store a mean vector for each cluster and its assigned label, then, for any new document, we compare its vector to the mean vector corresponding to each label. The one that is most similar is assigned the new label.

We know that the clustering algorithm was applied to maximize similarity to the mean. That suggests using the same classifier for new data. In general, using just the means of a set of measurements for prediction is a very weak method for most prediction problems. A small number of classes with a high variance would suggest that using means is highly simplistic. A large number of classes with few examples in each class might suggest poor generality for assigning new documents. These issues increase the importance of specifying the right number of classes. Do these weaknesses imply that k-means is inadequate? Not at all. We start with no labels. We have the freedom to choose our path unconstrained by a set of correct answers. Why not choose a straightforward and efficient approach from among the multitude of clustering methods developed by researchers in the absence of strict guidelines on correctness?

The accompanying software includes an implementation of k-means for documents. The k-means clustering may well be the world's most widely applied method for statistical clustering. It is readily adapted to documents and the vector representation, but it is also widely used in other fields.

5.2.1.1 Centroid Classifier

The mean of document vectors in a category is often referred to as the centroid of the category. It can be used for the purpose of prediction. Assume that we have k categories indexed by $c = 1, \ldots, k$, each with a centroid m_c. When a future document x comes in, their inner product $x \cdot m_c$ can be computed, and we pick the category associated with the largest inner product as the label of the document.

Class	Class Vectors	Normalized Centroids (m_{class})
$C1$	(1,2,0,0), (0,2,0,0)	(.24,.97,0,0)
$C2$	(1,0,1,1), (0,0,1,2)	(.27,0,.53,.80)

New document to be classified: (1,0,1,0)

Normalized new document (x): (.71,0,.71,0)

Similarity (inner product) between x and m_{C1}: .17

Similarity (inner product) between x and m_{C2}: .57

New Document classified as: $C2$

Fig. 5.9 An example of centroid classification

Figure 5.9 shows an example with four training documents partitioned into two classes, $C1$ and $C2$. The new document in this example is classified as $C2$ by this method.

The classification procedure described above is called the *centroid method*. If the documents and the centroid vector are normalized, then the inner product corresponds to the cosine measure. The centroid method has an intuitive meaning in this case since the closer the two documents are, the larger their inner product will be. The value of their inner product is a quantity in [0, 1], which can be used to measure the confidence of the classification result.

Although, generally speaking, the prediction accuracy of the centroid method is lower than for other prediction methods described in Chap. 3, it has certain advantages that can be important in practice. First, computation of the centroid is very simple and efficient. One only needs to store the centroid vector. The method can be easily adapted to the online setting since we only need to store the total sum and the number of vectors in order to compute the centroid. Secondly, the cosine measure has a very intuitive meaning, which can be used as an indicative measure to determine the confidence of the classification result.

In addition, the construction of the centroid of a category only requires positive data in the category and does not depend on negative data outside of the category. This is a very desirable property that can be important in some practical applications. For example, a dynamical environment often requires the users to constantly update the category structure and either include or remove different types of documents. The whole document collection changes frequently, although the definition of any particular topic may remain the same over time. Since in this case the distribution of the negative data changes constantly, a classification method that relies only on the positive data becomes very desirable. Next we turn our attention to another method that is particularly attuned to information retrieval.

5.2.2 Hierarchical Clustering

Hierarchical (agglomerative) clustering is a popular alternative to k-means clustering of documents. As expected, the method produces clusters, but they are organized

Fig. 5.10 The hierarchical clustering algorithm

1. Start with all documents as single-member clusters.
2. Find the best pair of clusters having no parents: B and C.
3. Combine the documents in B and C into a parent cluster A.
4. If more than one cluster remains with no parents, go to step 2.

in a hierarchy much like a table of contents for a book. Similarity is decided by averages and also by maximum and minimum distances between documents within a cluster, measures that some feel are more consistent with information retrieval and its similarity measures.

The major weaknesses of hierarchical clustering are timing and computational complexity. The method is intended for clustering a relatively small number of documents. For k-means, we expect the number of clusters to be far smaller than the number of documents. Each comparison is with the cluster mean, so the computations and comparisons grow approximately linearly with the number of documents. For hierarchical clustering algorithms that compare pairs of documents and not averages, the similarity between all pairs of documents is computed prior to clustering, so the complexity is of the order of the number of documents squared. The algorithms that involve using the least similar document in a cluster as the best are even more computation intensive. Thus we need to be judicious in applying these methods.

Figure 5.10 is an algorithm for hierarchical agglomerative clustering (HAC). The result will be a binary tree with links of parent clusters to their partitioned clusters. The root node contains all documents, and these are recursively partitioned into smaller clusters. Unlike decision tree induction for prediction, the HAC trees are built from the bottom up. Pairs of clusters are recursively combined until only one cluster remains. The terminal nodes of the tree are the actual clusters. A tree may be pruned to find a smaller tree that is more desirable than the fully grown tree.

Figure 5.11 is an example of hierarchical clustering. The same data used in Fig. 5.8 to illustrate the k-means algorithm are used here as well. If we specify a fixed number of clusters, in this example two clusters, then we see that k-means performs somewhat better. To get two clusters, we prune the decision tree to just the root node and its two children. Comparing the variance, the average squared error from the cluster mean shows a higher variance for hierarchical clustering. If a hierarchy or taxonomy is not needed for the application, and an average value is sufficient to characterize a cluster, then k-means has the advantage. Using either single or complete link analysis may be a better fit to information retrieval, where the k-nearest neighbors are of interest. Using the single-link criterion (i.e., most similar document) can lead to chaining documents that are distant to other members. The complete-link analysis (i.e. least similar document) is considered more desirable, but has an extravagant computational time.

The binary tree produced by the hierarchical clustering method is a map of many potential groupings of clusters. We can process this map to get an appropriate number of clusters, something that is more difficult with k-means, where we usually

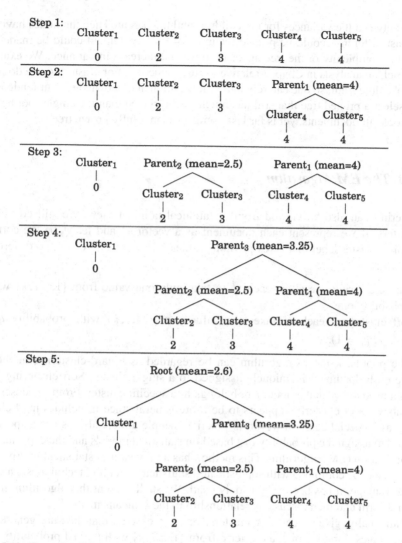

Fig. 5.11 Example of hierarchical clustering

rerun the procedure when we specify a new value of k. If we have a specific number of clusters in mind, then the tree can be pruned at the right depth to ensure that number of clusters. We followed that route in our example. The leaf nodes of the pruned tree represent the accepted clusters.

Chopping a tree to produce equal-length branches is one way to get a result corresponding to a fixed number of clusters. It also produces a balanced table of contents. It is often unnatural for an application when the tree produces a lengthy chain of entries. From a prediction perspective, one is tempted to prune differently. Prune the branches that result in the least change to some overall measure of performance. In the case of using a mean vector as a summary of the cluster, a comparison could be made between the variances for the children and the parent. Then the nodes having the least difference could be pruned recursively and a judgment could be made to trade off complexity of the tree against a potential increase in variance. We examined such an analysis in Chap. 3 relative to rule induction. For clustering, we do not have absolute truth, so one can rely more on intuition and knowledge. Our tendency is to select a pruned tree that still has much less variance than the single root node that pools all documents, yet is far less complex than a fully grown tree.

5.2.3 The EM Algorithm

Clustering can also be viewed from a statistical point of view. We still consider k-clusters. If we represent each document as a vector x^i and let y^i be the corresponding cluster label that takes possible values $\{1, \ldots, k\}$, there are two different scenarios:

- Hard clustering: we allow each y^i to take only one value from $\{1, \ldots, k\}$ with probability one.
- Soft clustering: each y^i takes the value $c \in \{1, \ldots, k\}$ with probability $q_{i,c}$ ($\sum_{c=1}^{k} q_{i,c} = 1$).

The popular k-means algorithm can be regarded as a hard-clustering method, where each document is uniquely assigned to a single cluster. Soft clustering assigns a probability that a document belongs to a specific cluster. From the description above, soft clustering appears to be more general since it includes hard clustering as a special case. However, in practice, people almost always use a specific method to assign the probability $q_{i,c}$ based on statistical models and the expectation maximization (EM) algorithm. This method has a very natural statistical interpretation and can be combined with sophisticated statistical modeling techniques. Due to this advantage, it has been very widely used. We shall present this algorithm in its general form and then discuss its relationship to the k-means method.

Statistically, given a set of k clusters, one can assume that data are generated in two stages. First we pick a cluster c from $\{1, \ldots, k\}$ with a fixed probability μ_c ($\sum_{c=1}^{k} \mu_c = 1$); then we generate data points x according to a probability distribution $p_c(x|\theta_c)$. The model parameters θ_c for each cluster c should be estimated from the data. They characterize the unknown information that we want to determine from the data with our clustering algorithm.

Mathematically, the data generation mechanism above corresponds to a statistical model called *mixture model*, where the likelihood of a data point x is

$$p(x) = \sum_{c=1}^{k} \mu_c p_c(x|\theta_c). \tag{5.2}$$

Each cluster c is a mixture component. Given a data point x, its label y can be interpreted as the cluster that generates x. The probability of x being generated from the cluster $\ell \in \{1, \ldots, k\}$ is given by

$$p(y = \ell|x) = \frac{\mu_\ell p_\ell(x|\theta_y)}{\sum_{c=1}^k \mu_c p_c(x|\theta_c)}. \tag{5.3}$$

Here y represents the hidden label of x, representing the actual cluster that x belongs to. This is a natural formula to use for assigning cluster labels in a statistical soft clustering algorithm and is used in the EM method.

The mathematical derivation of EM is based on an important property of (5.3), which we will consider next. For each data point x, we may introduce hidden variables q_y for each label $y = 1, \ldots, k$ such that $[q_y]$ is a probability measure: $q_y \geq 0$ and $\sum_{c=1}^k q_c = 1$. The meaning of each variable q_c is the conditional probability that x belongs to cluster c (that is q_c represents $p(y = c|x)$). A well-known fact from information theory states that the mutual entropy between $[q_c]$ and $[p(y|x)]$, defined as $\sum_{c=1}^k q_c \ln(q_c/p(y = c|x))$, is always nonnegative. It is easy to see that the equality with zero can be achieved at $q_c = p(y = c|x)$. Using the expression of $p(y = c|x)$ in (5.3), we can rewrite this inequality as

$$\ln \sum_{c=1}^k \mu_c p_c(x|\theta_c) \geq \sum_{c=1}^k q_c \ln \frac{\mu_c p_c(x|\theta_c)}{q_c},$$

where the equality holds with $q_c = p(y = c|x)$. This property, which we restate below in (5.4), is a crucial step in the derivation of the EM method.

$$\ln \sum_{c=1}^k \mu_c p_c(x|\theta_c) = \max_{q_1,\ldots,q_k} \sum_{c=1}^k q_c \ln \frac{\mu_c p_c(x|\theta_c)}{q_c}, \quad q_c \geq 0, \sum_{c=1}^k q_c = 1, \tag{5.4}$$

where the maximum on the right-hand side is achieved at $q_c = p(y = c|x)$ with $p(y|x)$ given by (5.3).

Given a mixture model as in (5.2), the most widely used statistical estimation technique for finding the parameters θ_c ($c = 1, \ldots, k$) from a set of data x^i is the so-called *maximum-likelihood estimation method*, which picks a parameter to maximize the observed likelihood of the data $\prod_{i=1}^n p(x^i)$. The formula to compute parameters in naive-Bayes, such as (3.7), can also be derived from the maximum-likelihood method. For clustering, we find θ_c by solving the following optimization problem:

$$\max_{\theta_1,\ldots,\theta_k} \sum_{i=1}^n \ln \sum_{c=1}^k \mu_c p_c(x^i|\theta_c). \tag{5.5}$$

The parameters μ_c are assumed to be given and often just take a uniform distribution $\mu_c = 1/k$. The EM method is essentially a numerical technique to find a local optimum of (5.5). It is based on the following reformulation of (5.5) using (5.4):

$$\max_{\theta_1,\ldots,\theta_k} \sum_{i=1}^n \max_{q_{i,1},\ldots,q_{i,k}} \sum_{c=1}^k q_{i,c} \ln \frac{\mu_c p_c(x^i|\theta_c)}{q_{i,c}}, \quad q_{i,c} \geq 0, \sum_{c=1}^k q_{i,c} = 1. \tag{5.6}$$

$$
\begin{aligned}
&\textbf{Initialize } \theta_1, \ldots, \theta_k \\
&\textbf{iterate} \\
&\quad \textbf{for } i = 1, \ldots, n \textbf{ do} \\
&\qquad q_{i,c} = \frac{\mu_c p_c(x^i | \theta_c)}{\sum_{l=1}^{k} \mu_l p_l(x^i | \theta_l)} \quad (c = 1, \ldots, k) \quad (\text{E step}) \\
&\quad \textbf{end for} \\
&\quad \textbf{for } c = 1, \ldots, k \textbf{ do} \\
&\qquad \text{Solve } \theta_c \text{ by solving } \max_{\theta_c} \sum_{i=1}^{n} q_{i,c} \ln p_c(x^i | \theta_c) \quad (\text{M step}) \\
&\quad \textbf{end for} \\
&\textbf{until } \text{convergence}
\end{aligned}
$$

Fig. 5.12 The EM algorithm

Here, the meaning of $q_{i,c}$ is the conditional probability that the true cluster label y^i of x^i is c. The EM algorithm is an alternating optimization method applied to (5.6). In the E step, we fix θ_c and solve for $q_{i,c}$. Since each subproblem has the same form of (5.4), the solution is given by (5.3). In the M step, we fix $q_{i,c}$ and solve for θ_c for each c. The method is summarized in Fig. 5.12.

Intuitively, the EM procedure may be explained in the following way. Although we do not know the true cluster label y^i for x^i, we can estimate the conditional probability of $y^i = \ell$ for each cluster $\ell = 1, \ldots, k$ using (5.3). This is the E-step. With this estimate, we can divide each data point x^i into k parts, each for a cluster ℓ, weighted by $p(y^i = \ell | x^i)$, the conditional probability of x^i belonging to cluster ℓ. Using this division, we can then estimate the cluster parameter θ_ℓ independently for each cluster ℓ (using the maximum likelihood method), using only the ℓ-th part of each training point x^i (i.e., weight each x^i by $p(y^i = \ell | x^i)$). This is the M-step.

Although in theory the EM method may converge slowly, practitioners often find that 20 iterations gives satisfactory results. It is also necessary to start EM with different initial parameters that are often generated randomly. This is to improve local optimal solutions found by the algorithm with each specific initial parameter configuration.

The EM algorithm for maximizing (5.6) is based on statistical modeling and maximum-likelihood estimates. By modifying (5.6), we can also obtain different clustering procedures (which do not necessarily have maximum-likelihood estimate interpretations). For example, hard-thresholding methods such as k-means can be obtained by changing (5.6) to

$$
\max_{\theta_1, \ldots, \theta_k} \sum_{i=1}^{n} \max_{q_{i,1}, \ldots, q_{i,k}} \sum_{c=1}^{k} q_{i,c} \ln(\mu_c p_c(x^i | \theta_c)), \quad q_{i,c} \geq 0, \sum_{c=1}^{k} q_{i,c} = 1. \quad (5.7)
$$

To solve this optimization problem, we can still use the EM algorithm in Fig. 5.12 but with the E step replaced by the hard cluster assignment rule: $q_{i,c} = 1$ if $c = \arg \max_l (\mu_l p_l(x^i | \theta_l))$, and $q_{i,c} = 0$ otherwise.

The EM algorithm can be used with any probability model, such as the naive Bayes model, which we described in Chap. 3. Another frequently used model is the Gaussian model, where we assume $p_c(x | \theta_c) \propto \exp(-\frac{(\theta_c - x)^2}{2\sigma^2})$. The simplest case is

Document Vectors	$x^1 = [0], x^2 = [1], x^3 = [2], x^4 = [4], x^5 = [4]$	
	Cluster 1	**Cluster 2**
Initial:	mean=2.0	mean=3.0
Step 1: E step	$q_{\cdot,1} = [0.9, 0.8, 0.6, 0.2, 0.2]$	$q_{\cdot,2} = [0.1, 0.2, 0.4, 0.8, 0.8]$
M step	mean=1.3	mean=3.3
Step 2: E step	$q_{\cdot,1} = [1.0, 0.9, 0.6, 0.0, 0.0]$	$q_{\cdot,2} = [0.0, 0.1, 0.4, 1.0, 1.0]$
M step	mean=0.9	mean=3.6
Step 3: E step	$q_{\cdot,1} = [1, 1.0, 0.7, 0.0, 0.0]$	$q_{\cdot,2} = [0.0, 0.0, 0.3, 1.0, 1.0]$
M step	mean=0.9	mean=3.7

Fig. 5.13 An example of EM clustering

to assume that σ is a known quantity such as 1, and $\mu_c = 1/k$. Under this assumption, Fig. 5.12 can be used to compute the mean vectors θ_c for the Gaussian mixture model, where the E and M steps are given by

- (E step) $q_{i,c} = \dfrac{\exp(-\frac{(x^i - \theta_c)^2}{2\sigma^2})}{\sum_{l=1}^{k} \exp(-\frac{(x^i - \theta_l)^2}{2\sigma^2})}$,
- (M step) $\theta_c = \dfrac{1}{\sum_{i=1}^{n} q_{i,c}} \sum_{i=1}^{n} q_{i,c} x^i$.

This clustering method can be regarded as a soft k-means method. If we replace the E step by the hard cluster assignment rule mentioned above, then we obtain the standard k-means method.

Figure 5.13 gives an example of EM iterations with a two-component Gaussian mixture model and $\sigma = 1$. The mean vectors are initialized to 2.0 and 3.0. We report the probability assignment $q_{i,c}$ for each document i in each cluster c after the E step and the mean-vectors for the clusters after the M step.

5.3 What Do a Cluster's Labels Mean?

Let's assume that we have successfully clustered the documents. At the end of the process, we have k clusters. We can assign numbers to correspond to different clusters. Documents within the same cluster are expected to be similar. Yet, we need greater insight into the process than just assigning a label without expressing meaning. In a labeled prediction problem, the labels are assigned with care, often corresponding to some exact event. For example, in text categorization, the labels are specific categories for indexing documents, such as sports or financial newswires. Although computer programs have no need for associating meanings to clusters, the reason for even attempting prediction is the meaning and significance of the list of labels. Where is this meaning for our clustering labels?

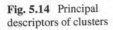

Fig. 5.14 Principal
descriptors of clusters

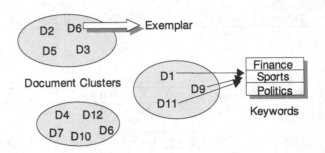

Figure 5.14 shows the principal descriptors of the automatically generated cluster labels. They are explained and motivated by key words or exemplar documents. We should not lose sight of what we are clustering. Documents are composed of words, and the distribution of words is the basis of document clustering. Surely we can use these words to express the meaning of the result of the clustering. Documents are clustered by similarity of words, so if we characterize a cluster by the right words, we should be able to give the meaning or rationale for the reasonableness of the cluster.

The clusters have been assembled by automated procedures, a task that humans would not attempt on their own. In a follow-up step, the computer may indicate the key words that are representative of each cluster. This step is not very different from composing a local dictionary for labeled data. In Chaps. 2 and 3, several techniques were described for local dictionary composition relative to a single label. For example, we could take all the basketball news stories in a collection and create a dictionary solely from those news stories. Similarly, we can compose a local dictionary for each cluster. The simplest approach is to take the most frequent words and then remove the stopwords. Alternatively, the words with the largest average tf-idf measures might be used. Another approach might involve feature selection procedures that select a relevant set of words from a larger set, either a global dictionary or a large local dictionary. The main concept is to find words that are representative of the cluster. The k-means clustering routine in the accompanying software uses a scheme based on tf-idf measures to describe each cluster using its most representative words. Instead of relying completely on automated procedures, the human expert can review the words generated by automated procedures and make a final determination.

A label for a cluster can summarize very large numbers of documents. A cluster's key words help give meaning to those labels. Because clustering procedures process unlabeled data, we might be particularly demanding in exploring the value of their results. Unlike data for typical numerical applications, our documents have meaning in their text. Someone has carefully written and published a document. Why not use one or more documents to characterize the cluster? The technical problem is to find a small number of documents to serve as exemplars for the much larger group of documents in a cluster. There are many ways to do this. One simple approach is to select the document that is most similar to the cluster mean vector. If the cluster spans a large number of words, additional exemplars might be retrieved. In addition to comparing potential exemplars with the mean vector, a pair of least similar documents might also be considered. The procedures for selecting exemplars are far from perfect. We expect the documents in a cluster to be similar. Thus, we should be able to select a few that are "typical" documents. The human expert can read and review these retrieved documents to understand the results of the clustering process and to reach some decisions about their value.

5.4 Applications

We have presented clustering as the assignment of labels to documents. We saw this within the context of a problem with the spreadsheet that we use to represent data for prediction—it was missing the last column, and now we have methods to compute and fill in the last column. In that sense, our venture into clustering has been successful. Labels are necessary for the application of prediction methods. Clustering could have very wide applicability because it is often expensive to get correct labels. Surely we would like to avoid that expense. In practice, clustering is a weaker process than learning from labeled data, and the expense is often justified by more accurate predictions.

Beyond assigning labels, clustering has natural capabilities that are especially beneficial for exploratory analysis. The hierarchical methods give structure to the data, producing a taxonomy that maps relationships among the documents. These are ways of summarizing and organizing documents whose connections are poorly understood. The problems reported to help desks are good examples of the potential benefits of clustering. Document collections are digital libraries, and we could make an analogy with a traditional paper-based library. Clustering is a conceptual way to organize the library, placing related books, such as mysteries, in a common location. In our case, the clustering methods determine which books are related and also determine the topics and hierarchical relationships among the topics. In a modern paper-based library, search by computer has somewhat eclipsed the classical organization by topic. Users can search dynamically, and when they find the right material, they just need a pointer to a shelf, or even a robot, that will retrieve the selected item. One might argue that, for many applications, clustering has also been eclipsed by search engines that can dynamically find nearest neighbors and allow users to interact and reformulate their queries.

One view of the benefits of clustering is the summarization of properties of a document collection. To the previous types of summaries, we can add redundancy and frequency detection. Using similarity measures, it is easy to find repeats of documents. They match in every word. By changing the threshold of matching words, documents that are potentially very similar, such as a problem report on the same form of breakdown, can be detected. That gives us the opportunity to count the number of similar items in a cluster or perhaps delete documents to eliminate redundancy.

Summaries of clusters are clearly of value in exploratory reviews of document collections. Our main task is prediction. Although labels can be generated by clustering, if they are not accurate, we cannot move closer to reaching our primary goal.

Next we look at evaluating the results of clustering, which give clues to the overall success of both the clustering and prediction efforts.

5.5 Evaluation of Performance

For prediction problems, evaluation is straightforward. The computed answers are compared with the correct answers, and accuracy or error rates are determined. If we want to be fair and unbiased, we hide documents and only at the very end test our methods and compare their results with the correct answers. Clustering starts with unlabeled data, and using the same approach to evaluation is not feasible.

Let's not worry about projecting into the future. Let's just consider the quality of the clustering results. The concept of clustering is to put similar documents in the same group. One way of evaluating whether this has been achieved is to compute a cluster mean and its variance or standard deviation. We can consider a deviation from the mean to be a form of error and the standard deviation to be the typical error. If documents within a cluster are similar, the variance of the mean will be low. In (5.8) the overall measure of variance for all documents is computed, where $S(i)$ is the similarity result for document i, and $MS(i)$ is the average of the similarity measures for all documents in the same cluster as document i. The variance can be computed over all n documents in the collection to get a global indication of effectiveness. It can also be computed for an individual cluster to determine its performance and to compare it with other clusters.

$$\text{Variance} = \sum_i \frac{(S(i) - MS(i))^2}{n},$$

$$(5.8)$$

$$\text{Standard Deviation} = \text{Error} = \sqrt{\text{Variance}}.$$

How is the similarity measure $S(i)$ of a document computed? Similarity is computed between a pair of documents. We have mentioned many choices for similarity measures, any of which could be used in (5.8). For example, similarity can be computed using normalized tf-idf and cosine similarity. For k-means, a comparison would be made between each document vector and the mean vector. Alternatively, we might simply count shared words. For k-means, each document would be compared with the mean vector to see the number of shared words. In (5.8), $S(i)$ is the number of words shared between document i and the mean vector of its cluster, and the average of all $S(i)$ in the same cluster is $MS(i)$. As was done in hierarchical clustering, $S(i)$ could be similarity between document i and its most similar document in its cluster or its most dissimilar document. Whatever measures are used to compute similarity, they are averaged and a variance is computed. If we were using just simple shared word counts, then a mean would indicate the average number of words shared between documents. Error could be measured by the expected deviation from this number. The error obtained by assigning all documents to the same single cluster

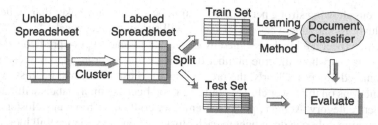

Fig. 5.15 Prediction and evaluation of unlabeled data

is the baseline measure which must be improved greatly to demonstrate meaningful results.

Lurking in the background is the number of clusters and their size. It would be very easy to specify a very large number of clusters, each containing few documents. The variance computed for each cluster will be low, but the usefulness of the cluster may also be poor. The tradeoff of complexity and error is implicit in this process. Without correct labels, one has to have design goals and knowledge of the application. The approximate number of clusters could be known, or an upper bound on an acceptable variance might be declared so that a relatively small number of clusters are selected.

These measures of clustering performance can be quite informative. They summarize the overall quality of the clustering process and can pinpoint clusters that differ in performance from others. Taken with the other forms of summarization, such as key words and extracts, the clustering results can help people make informed decisions and interpretations. Although a fully automated process can be specified for document clustering, it is wise to use knowledge to enhance the final results. We all read and comprehend documents, so we can examine the clustering results and evaluate the results on our own. One might even make a suggestion, such as altering the number of clusters or the measures of similarity to predict the automated process. One should not dismiss subjective human evaluation.

Document clustering is very compatible with information retrieval because it organizes a static collection of documents. Our task is prediction, generalizing to new unseen documents. Can we predict and then evaluate without labels? Figure 5.15 is an outline of how we might accomplish this mission. Let's look at the mechanics of prediction. We have a spreadsheet of structured data. The clustering process has provided a column of labels, so our spreadsheet is now complete and ready for the application of predictive learning methods. In Chap. 3, we described the appropriate way to organize data for prediction. Separate the data into train and test sets, which is typically done by assigning the most recent documents to the test set. The learning methods are applied to the training data, and the test data are used for evaluation. Error analysis is the basis for evaluating various error rates. Recall and precision can be measured. For the labels found by clustering, let us perform those same forms of evaluation. It is true that the cluster methods look at all of our data to assign labels. Our results for prediction will be biased optimistically because the clustering methods viewed both train and test data. We should accept that weakness and move on.

Evaluations showing strong performance might be somewhat weaker than indicated, but evaluations showing weak performance will surely be very weak.

Can we expect the labels assigned by clustering to lead to poor prediction? Consider a situation where the true number of labels is much larger than the results for the automated process. Clearly, the predictions for these labels will be missed. Yet, the predictions on the fewer classes could be good because many labels will fall into the "other" category. At the opposite extreme, we could have too many clusters, perhaps assigning only one document to each cluster. Then the variance will look great, but a nearest-neighbors method would have terrible results. To get good predictive results, both the cluster and prediction methods must be successfully applied. That does not guarantee that predictions are made for the most desirable outcomes. Clustering is empirical, but it also projects with poetic license, having the potential for going off in the wrong direction. However, if the labels appear reasonable, then we may get some useful predictions without the expense of human-assigned labels.

5.6 Summary

Document collections are frequently encountered without labels. Labels may be determined by clustering the documents into disparate groups and implicitly finding common themes among the document clusters. This chapter describes methods for clustering documents. A key theme for document clustering is computing measures of similarity. We review the major clustering methods: k-means clustering, hierarchical clustering and the EM algorithm. Strategies for assigning meaning to algorithmically generated clusters and labels are considered. Performance evaluation helps determine the empirical characteristics of desirable clusters.

5.7 Historical and Bibliographical Remarks

Clustering has been a topic of study in information retrieval for many years. Rocchio's method for clustering (Salton 1971) bears a close resemblance to k-means. Instead of allocating the document set to k bins, the Rocchio method picks k documents as exemplars and then finds which of the remaining documents are "close to" the exemplars. A special class is created for documents close to none of the exemplars. At each iteration, the cluster centroids are recomputed and the documents are reallocated until there is no longer any significant change in cluster membership.

Clustering was first applied to both terms and documents, but experiments with the SMART system did not show any advantage to clustering terms, and this activity was more or less abandoned (but see Bekkerman *et al.* 2003). The continued interest in clustering is partly due to the cluster hypothesis: "the associations between documents convey information about the relevance of documents to queries" (Jardine and van Rijsbergen 1971). The question is, whether document collections satisfy the cluster hypothesis. Answers to this question are not entirely clear-cut (Voorhees

1985). Nonetheless, there has been continued interest in clustering, particularly hierarchical agglomerative clustering methods. A good review of this literature is in Willett (1988).

More recently, clustering has been applied to World Wide Web searches. Queries to the Web often result in very many documents, and it is hoped that organizing the voluminous output will make it more useful (Dumais and Chen 2000). Another approach is based on the idea that many Web searches are very general and exploratory. A method for browsing these amorphous search results, called "scatter/gather," was proposed in Cutting *et al.* (1992). Search output is clustered, the clusters are presented to a user, who selects a few clusters as interesting, the documents in these clusters are reclustered, and the new, presumably more specific clusters are presented again.

Finally, clustering has been used in studies of multidocument summarization (Stein *et al.* 2000). Each document is summarized, the summaries are clustered, a passage from each cluster is selected for the final summary, and the passages are reordered.

The statistical approach of clustering based on mixture models has a long history. Although it is possible to use other optimization techniques to solve mixture models, the most widely used method is the EM algorithm due to its simplicity and intuitive statistical interpretation. The method was first presented in full generality in Dempster *et al.* (1977). It has since attracted enormous attention. Many works have appeared both on the application side and on the theoretical analysis of the method. In engineering applications, the EM algorithm has often been used with Gaussian mixture models. However, in text processing, it can be used with the multinomial naive Bayes model (Nigam *et al.* 2000) since it is more natural to assume that each cluster generates text through a naive Bayes model.

The technique of k-means clustering appeared in MacQueen (1967). It has since been used widely in practice. It is also simpler to implement and computationally more efficient than the EM method. For text clustering, a version of the k-means algorithm was studied in Dhillon and Modha (2001). Improvements to k-means are described (Elkan 2003). Properties of the more general hard-clustering method presented in (5.7) were considered in Kearns *et al.* (1997). A survey of statistical model-based clustering methods with experiments on text data can be found in Zhong and Ghosh (2003). Evaluation criteria for document clustering methods and their application to prediction are discussed in Banerjee and Langford (2004).

5.8 Questions and Exercises

1. Explain why Euclidean distance is not an appropriate metric for document similarity.
2. Verify the computations in Fig. 5.13 using the algorithm in Fig. 5.12.
3. How can the EM algorithm be used to perform a k-means clustering?
4. List any one advantage of using the EM algorithm over the k-means algorithm.

5. In k-means, the means of bins change every time a document is moved. How can the bin means be recomputed efficiently?

6. How is the centroid classifier related to nearest neighbour methods?

7. If we have a set of documents that are mostly similar except for a number of important exceptions, which clustering method would be best to extract out the primary cluster?

8. Show how the equations on p. 105 are obtained from (5.6).

9. How would you use a rule-based classifier to obtain useful descriptions of a set of clusters?

10. Given a set of document clusters, devise a scheme for labeling the clusters so as to distinguish them from each other. A label should also represent the characteristics of the associated cluster.

11. Use the k-means program in TMSK to generate 10 clusters of the Reuters training data. Re-run to generate 20 clusters, then 30 clusters. Discuss the sorts of clusters you get and any hints they might provide as to the optimal k for this document collection.

12. Explore the software further. Most of the examples are with the Reuters data. Try to download new datasets from the web and analyze them in a similar manner.

13. See the website http://clusty.com for a good example of clustering computing in text mining. Spend some time trying this search engine (it generates clusters based on the web pages retrieved by a search) and compare the results with that of your usual search engine.

Chapter 6
Looking for Information in Documents

An important research area for natural language processing and text mining is the extraction and formatting of information from unstructured text. One can think of the end goal of information extraction in terms of filling templates codifying the extracted information. In this chapter, we shall describe information extraction from this perspective and some machine-learning methods that can be used to solve this problem.

6.1 Goals of Information Extraction

Natural language text in digital form is a very important source of information. Such information is presented in an unstructured format that is not immediately suitable for automatic analysis by a computer. We still are a long way from achieving complete computer understanding of natural language text. However, computers can be used to sift through a large amount of text and extract restricted forms of useful information, which can be represented in a tabular format. Therefore, information extraction can be regarded as a restricted form of full natural language understanding, where we know in advance what kind of semantic information we are looking for. The main task is then to extract parts of text to fill in slots in a predefined template.

We illustrate the concept of information extraction using examples of identifying executive position changes. Every day there are many news articles and announcements containing information on executive position changes in different companies. Such information can be very useful for business intelligence but is represented in an unstructured form not suitable for automatic computer processing. For example, we do not know where in the text to find a person or the organization to which he or she belongs. In order to represent such information, a computer program is needed to automatically annotate the text. Consider the text given in Fig. 6.1, where entities that we are interested in are denoted by italic fonts. The text is taken from an article in The Wall Street Journal on February 24, 1994, and the named entities were annotated in the Sixth Message Understanding Conference (MUC-6). The extracted

S.M. Weiss et al., *Fundamentals of Predictive Text Mining*,
Texts in Computer Science 41,
DOI 10.1007/978-1-84996-226-1_6, © Springer-Verlag London Limited 2010

One of the many differences between *Robert L. James, chairman and chief executive officer of McCann-Erickson,* and *John J. Dooner, Jr.,* the agency's president and chief operating officer, is quite telling: Mr. James enjoys sailboating, while Mr. Dooner owns a powerboat.

Now, Mr. James is preparing to sail into the sunset, and Mr. Dooner is poised to rev up the engines to guide *Interpublic Group's McCann-Erickson* into the 21st century. Yesterday, *McCann* made official what had been widely anticipated: *Mr. James, 57* years old, is stepping down as chief executive officer on *July 1* and will retire as chairman at the *end of the year.* He will be succeeded by *Mr. Dooner, 45 ...*

Fig. 6.1 WSJ text with entity mentions emphasized by italic fonts

Table 6.1 Extracted position change information

Organization	*McCann-Erickson*
Position	*Chief executive officer*
Date	*July 1*
Outgoing person name	*Robert L. James*
Outgoing person age	*57*
Incoming person name	*John J. Dooner, Jr.*
Incoming person age	*45*

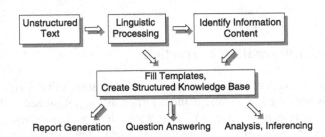

Fig. 6.2 Information extraction system

information is represented in Table 6.1 and can be stored in a relational database for additional processing and analysis.

The task of information extraction naturally decomposes into a sequence of processing steps, typically including tokenization, sentence segmentation, part-of-speech assignment, named entity identification, phrasal parsing, sentential parsing, semantic interpretation, discourse interpretation, template filling, and merging. A general information extraction system is illustrated in Fig. 6.2. Although the most accurate information extraction systems often involve handcrafted language-processing modules, substantial progress has been made in applying predictive machine-learning techniques to a number of processes necessary for extracting information from text.

The application of machine-learning techniques to information extraction is motivated by the time-consuming process needed to handcraft these systems. They require special expertise in linguistics and additional skills in artificial intelligence and

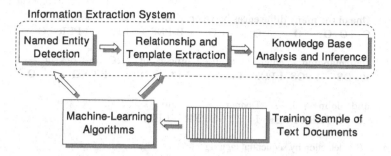

Fig. 6.3 Learnable information extraction system

computational linguistics. This, coupled with the expectation that domain-specific information is required to build an accurate system, motivates research and development of trainable information extraction systems.

The general architecture of a machine-learning-based information extraction system is given in Fig. 6.3. There are typically two main modules involved in such a system. The purpose of the first module is to annotate the text document and find portions of the text that interest us. For example, in the first sentence of the text in Fig. 6.1, we want to identify the string *Robert L. James* as a person and the string *McCann-Erickson* as an organization. Once such entity mentions are extracted, another module is invoked to extract high-level information based on the entity mentions. In the example of Fig. 6.1, we want to identify that the person *Robert L. James* belongs to the organization *McCann-Erickson*. This information is then filled into slots of a predefined template.

6.2 Finding Patterns and Entities from Text

Typical information extraction systems only partially parse the text (i.e., chunk the text into typed phrases to the extent required to instantiate the attributes of a template). The key rationale for this approach is the assumption that, for a wide range of constrained applications, the predefined attributes of templates and the semantic types of the attribute values permit the system to make a reasonably accurate determination of the roles of the various typed chunks identified by the linguistic processors (named entity identifiers, noun phrase chunkers, verbal group chunkers, etc.) without attempting a complete sentence parse. The systems are typically organized as a cascade of linguistic processors.

The first, and usually the most important, aspect of information extraction is to find entity mentions from text. We shall describe the machine-learning approach to this problem in greater detail.

One	of	the	many	differences	between	Robert	L.	James	,
O	O	O	O	O	O	B-PER	I-PER	I-PER	O

chairman	and	chief	executive	officer	of	McCann-Erickson	,
B-POS	I-POS	I-POS	I-POS	I-POS	O	B-ORG	O

and	John	J.	Dooner	,	Jr.	,	...
O	B-PER	I-PER	I-PER	I-PER	I-PER	O	

Fig. 6.4 Entity detection by sequential tagging

6.2.1 Entity Extraction as Sequential Tagging

One way of looking at entity extraction is to annotate chunks of text strings with some prespecified types. For example, we want to find mentions of people, locations, and organizations in text. Our task is to divide text into syntactically related nonoverlapping groups of words (chunks). For example, the sentence *One of the many differences between Robert L. James, chairman and chief executive officer of McCann-Erickson, and John J. Dooner, Jr., the agency's president and chief operating officer, is quite telling* ... can be annotated as follows:

> One of the many differences between [PER Robert L. James], [POS chairman and chief executive officer] of [ORG McCann-Erickson], and [PER John J. Dooner, Jr.], the agency's [POS president and chief operating officer], is quite telling ...

In the example above, chunks that start with PER denote person, chunks that start with POS denote position, and chunks that start with ORG denote organization.

The most successful machine-learning-based approach to this task regards the problem as a token-based tagging problem. The idea is to divide text into tokens (words) and then assign each token a tag value that encodes the chunking information. There are many different encoding schemes. One commonly used method is to represent chunks by the following three types of tags:

B-X: first word of a chunk of type X,
I-X: noninitial word in a chunk of type X,
O: word outside of any chunk.

As an example, the sentence considered before in this section can be tokenized and annotated as shown in Fig. 6.4.

One can now view entity detection as a sequential prediction problem, where we predict the class label associated with every token in a sequence of tokens. In our case, each token is either a word or punctuation in the text. The advantage of this approach is that the task now becomes a simpler classification method, where the goal is to predict the class label associated with every token. Therefore, methods used for classification, such as those described in Chap. 3, can be applied.

6.2.2 Tag Prediction as Classification

In order to determine the label of a token, we create a feature vector for this token. The label is then determined by the feature vectors. In many statistical approaches, this requires us to estimate a conditional probability model. We use $x = [x_1, \ldots, x_d]$ to denote a d-dimensional feature vector associated with a token and t to denote the tag value of the token. The task is to estimate the conditional probability $\Pr(t|x)$; that is, the odds of a token's label being t if its associated feature vector is x. We can then assign the token a label that has the largest conditional probability score $\Pr(t|x)$.

If the number of possible labels and the number of feature values are very small, then one can form a table of all possible feature values. At each feature value, we compute the associated counts for every possible tag value. Given a feature value, we then simply predict the class label associated with the highest count. This approach does not work in practice since, in order to obtain good performance, it is necessary to use a very large set of features. Although we construct each feature to capture a particular linguistic pattern that is expected to help solve the problem, it is often quite difficult for us to tell how useful each feature actually is. The learning algorithm should have the ability to utilize a large set of features. The simple counting method fails for this purpose for two reasons: it is computationally infeasible to store the large table necessary for a large number of features, and we run into the so-called data sparsity problem in that we do not observe enough data to obtain accurate class label counts for each feature value.

A feature vector can be constructed in different ways. The most frequently used feature-encoding scheme is binary encoding, where each feature component takes a binary value: $x_j \in \{0, 1\}$. The value of $x_j = 0$ means that the feature component is not active, and $x_j = 1$ means that the feature component is active. Therefore, each feature component can be regarded as a test that asserts whether a certain pattern occurs or not. As an example, we can use a feature component to test whether the previous token is the word *Professor*, which is clearly a good indicator of whether the current token is in the B-PERSON category or not. In the binary-encoding scheme, the feature component is 1 (active) if the previous word is *Professor* and is 0 (inactive) if it is not.

Note that restricting ourselves to binary encoding does not impose any serious limitation. In fact, one may always encode a feature vector with multivalued feature components into a binary-valued feature vector since we can simply introduce additional feature components in the binary-valued feature vector, one for each multivalue that may occur. Although we can use a large set of features (possible tests of patterns) for information extraction, at any particular token, only a very small subset of features are active. Therefore, an additional requirement for a desirable classification method is the ability to handle sparse data efficiently.

The prediction methods described in Chap. 3 can be used for token-based tagging problems. Similar to text categorization, linear models are very successful for this task due to their ability to take advantage of large feature sets. This means that the designer does not have to worry about which feature is more important or more useful, and the job is left to the learning algorithm to assign appropriate weights to the corresponding feature components. A good learning algorithm should learn a linear classifier such that the importance of each feature component is reflected by its corresponding weight value. As mentioned in Chap. 3, in this approach, non-linearity can be explicitly captured by sophisticated nonlinear features. For tagging problems, each feature usually corresponds to a certain linguistic pattern that is predictive for a certain class. We shall describe some useful features in greater detail later.

Now assume that we are given a carefully constructed feature vector x, and our goal is to determine whether the label associated with x is t. In a linear model, we seek a weight vector w^t and a bias b^t such that $w^t \cdot x + b^t < 0$ if its label is not t and $w^t \cdot x + b^t \geq 0$ otherwise. In the multiple label case, we compute the score $s^t = w^t \cdot x + b^t$ for each possible value t and assign the label to the value t with the largest score s^t. An implementation of this is included in the accompanying software.

In Chap. 3, we described two methods that can be used to learn the weight vector for such a linear decision rule. One is naive Bayes, which is very simple and efficient but usually does not provide the best classification performance. The other method is a linear scoring method based on minimizing a robust loss function, which we call RRM (robust risk minimization). Both gave estimates of the conditional probability. Although the RRM method works very well for information extraction, an alternative method, called *maximum entropy*, has historically been used in the natural language processing literature. This method is related both to RRM and to naive Bayes but is computationally more expensive. Due to its importance in the information extraction literature, we shall describe it in detail.

6.2.3 The Maximum Entropy Method

Recall that in naive Bayes, we make the very simple assumption that, for each class t, each feature component is generated independently. Mathematically, (6.1) describes this probability model, where $\Pr(x_j = 1 | label = t) = e^{w^t_j}/(1 + e^{w^t_j})$ and $\Pr(x_j = 0 | label = t) = 1/(1 + e^{w^t_j})$.

$$\Pr(x|t) = \prod_{j=1}^{d} \Pr(x_j|t) = \frac{\exp(w^t \cdot x)}{\prod_{j=1}^{d}(1 + e^{w^t_j})}. \tag{6.1}$$

Let $c(w^t) = \prod_{j=1}^{d}(1 + e^{w^t_j})$ and $b^t = \ln \Pr(t) - \ln c(w^t)$, then we have the following $\Pr(x) = \sum_t \Pr(t) \Pr(x|t) = \sum_t \exp(w^t \cdot x + b^t)$. Using the Bayes rule, (6.2) expresses the conditional class probability

$$\Pr(t|x) = \Pr(t) \Pr(x|t)/\Pr(x) = \frac{\exp(w^t \cdot x + b^t)}{\sum_{t'} \exp(w^{t'} \cdot x + b^{t'})}. \tag{6.2}$$

Fig. 6.5 Graphical representation of generative and discriminative models

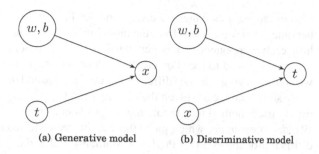

(a) Generative model   (b) Discriminative model

Since $\Pr(x)$ is independent of the class label t, we may assign each token with feature x a label t that has the largest score $s^t = w^t \cdot x + b^t$.

A major disadvantage of naive Bayes is that the conditional independence assumption in (6.1) is often too strong. This can be remedied by the maximum entropy (MaxEnt) method, which is one of the most popular and successful methods in natural language processing.

Observe that for classification problems we are interested in estimating the conditional probability $\Pr(t|x)$. The naive Bayes approach models the data probability $\Pr(x|t)$ for each class, which induces a conditional probability as in (6.2). In the literature, a classification method that models the class data probability $\Pr(x|t)$ is often called a *generative model*, while a method that directly models the condition probability $\Pr(t|x)$ is often called a *discriminative model*. Their graphical representations are given in Fig. 6.5, where the arrows represent the directions of the variable dependencies. Since x is often of very large dimensionality, it is usually much more difficult to have a good data model $\Pr(x|t)$ than a good conditional class probability model $\Pr(t|x)$. Therefore, the discriminative modeling approach in general outperforms the generative modeling approach.

The naive Bayes method can be considered as a method of estimating a conditional probability model of (6.2), but relies on the unrealistic independence assumption in (6.1). For a generative modeling approach such as naive Bayes, this kind of simplification is always necessary in order to prevent the data sparsity problem we mentioned earlier. It reflects the fact that estimating $\Pr(x|t)$ is often more difficult than estimating $\Pr(t|x)$ directly. Therefore, a simplification is required in the more difficult problem of estimating $\Pr(x|t)$, which leads to the relatively poor performance of naive Bayes.

The discussion above suggests that an immediate improvement of naive Bayes is to find weight vectors w^t that directly estimate the conditional probability model $\Pr(t|x)$ of (6.2) without going through the naive Bayes data model. A direct conditional probability model of (6.2) is often referred to as *maximum entropy* (MaxEnt) in computational linguistics. This is because an exponential model of this form can be obtained from the so-called maximum entropy principle. This particular relationship isn't important for our purposes, and thus we shall not explore it here. The same conditional model has also been widely used in statistics, and is often called (multicategory) *logistic regression*.

A significant advantage of the maximum entropy model over the naive Bayes model is that we do not need to have any independence assumption for the component of the features. Since we directly model $\Pr(t|x)$, how the feature x is generated becomes irrelevant. Therefore, the model in (6.2) does not make any assumption on how each component of x is generated. This also indicates that in MaxEnt modeling we only need to focus on creating features that are most useful for the problem, without worrying whether different features are redundant.

Assume that we are given the training data $(x^1, t^1), \ldots, (x^n, t^n)$. The most commonly used method to estimate the weight vector in (6.2) is through the *maximum-likelihood estimate*, which picks the weight vector to maximize the observed conditional probability $\prod_{i=1}^{n} \Pr(t^i|x^i)$. Mathematically, the method is equivalent to the minimization of

$$\min \sum_{i=1}^{n} \left[-(w^{t^i} \cdot x^i + b^{t^i}) + \ln \sum_{t=1}^{K} \exp(w^t \cdot x^i + b^t) \right]. \qquad (6.3)$$

The method above is similar to the RRM method described in Chap. 3 in that both seek weight vectors by minimizing loss functions averaged over the training data. The main difference is that they minimize different loss functions. In practice, performance of the classifier obtained is not very sensitive to different choices of loss functions. However, some formulations are numerically easier to solve than others.

One can observe that the model in (6.2) is unchanged when each weight w^t is increased by a constant vector. Therefore, the solution of (6.3) is not uniquely determined. One method to avoid this difficulty is to impose a constraint such as $\sum_t w^t = 0$ (and $\sum_t b^t = 0$). This is especially convenient for binary problems, where the labels take two values $\{\pm 1\}$. Another method of resolving this difficulty is to use regularization, such as what we have employed for the RRM method described in Chap. 3. However, practitioners of the MaxEnt model often are satisfied with finding any one of the possible solutions. A popular method for doing so is the so-called generalized iterative scaling (GIS) method.

Before describing the algorithm, we shall rewrite MaxEnt in a slightly different form. Consider weight vectors $w^1, \ldots, w^K$ and training data $x^1, \ldots, x^n$. For each $i = 1, \ldots, n$ and $t = 1, \ldots, K$, denote by $x^{i,t}$ a $(d+1)K$ dimensional vector, where its components from $(d+1)(t-1)+1$ to $(d+1)t-1$ are given by the corresponding components of x^i; its $(d+1)t$-th component is 1; and its other components are filled with zeros. That is, each $x^{i,t}$ has a form $x^{i,t} = [0, \ldots, 0, x^i, 1, 0, \ldots, 0]$. We also let w be a $(d+1)K$ dimensional vector, where for each $t = 1, \ldots, K$, its components from $(d+1)(t-1)+1$ to $(d+1)t-1$ is given by the corresponding components of w^t, and its component $(d+1)t$ is given by b^t. That is, w has the form $w = [w^1, b^1, w^2, b^2, \ldots, w^K, b^K]$. Using the notations above, the linear score $w^t \cdot x^i + b^t$ can be rewritten as: $w^t \cdot x^i + b^t = w \cdot x^{i,t}$. Therefore, the model in (6.2) can be regarded as a special case of the following slightly more general model, which often appears in the computational linguistics literature:

$$\Pr(t|x^i) = \frac{\exp(w \cdot x^{i,t})}{\sum_{t'} \exp(w \cdot x^{i,t'})}.$$

```
Let w = 0
Let r^{i,t} = 0 (i = 1, ..., n, t = 1, ..., K)
Let q_j = ∑_{i=1}^{n} x_j^{i,t^i} (j = 1, ..., (d+1)K)
Let q = max_{i,t} ∑_{j=1}^{(d+1)K} x_j^{i,t}
iterate
    p_j = 0 (j = 1, ..., (d+1)K)
    for i = 1, ..., n do
        z_i = ∑_{t=1}^{K} exp(r^{i,t})
        p_j = p_j + ∑_{t=1}^{K} x_j^{i,t} exp(r^{i,t})/z_i (j = 1, ..., (d+1)K)
    end for
    for j = 1, ..., (d+1)K do
        Δw_j = 1/q ln (q_j / p_j)
        w_j = w_j + Δw_j
        r^{i,t} = r^{i,t} + x_j^{i,t} Δw_j (i = 1, ..., n)
    end for
until convergence
```

Fig. 6.6 The GIS algorithm

The maximum-likelihood estimate in (6.3) can now be rewritten as

$$\min \sum_{i=1}^{n} \left[-w \cdot x^{i,t^i} + \ln \sum_{t=1}^{K} \exp(w \cdot x^{i,t}) \right]. \tag{6.4}$$

We now describe the GIS algorithm that solves (6.4). The algorithm is shown in Fig. 6.6. We use w_j to denote the j-th component of w and $x_j^{i,t}$ to denote the j-th component of $x^{i,t}$. We also assume that each $x_j^{i,t}$ is binary (0 or 1).

In GIS, $r^{i,t}$ maintains the current sum $w \cdot x^{i,t}$. The quantity q_j is the observed frequency of the j-th feature, and p_j is the expected frequency of the j-th feature under the maximum entropy model (with the current weight vector w). From the update rule of Δw_j, we can see that, if $q_j > p_j$, then w_j is increased, which has the effect of increasing the expected frequency p_j in the next iteration. Similarly, if $q_j < p_j$, then w_j is decreased, which has an effect of decreasing the expected p_j in the next iteration. Therefore, GIS tries to update the weight vector so that the expected frequency p_j becomes closer to the observed frequency q_j. It can be shown that p_j converges to q_j eventually. Note that a weight vector is the solution of (6.4) when $p_j = q_j$. This can be seen by setting the derivative with respect to each w_j to zero in (6.4).

Assume that we train a named entity recognition model using only tokens in a window of size ± 2. Table 6.2 shows an example of computed weights associated for some tokens with their positions relative to the current token. With these weights, consider the following sentence segment: *Russian Prime Minister Mikhail Kasyanov said on Monday* ..., and let us try to determine the label for word *Mikhail*. Although neither *Mikhail* nor *Kasyanov* appeared in the training data, the tagger correctly classified the tag associated with *Mikhail* to be B-PER based on

Table 6.2 Examples of
linear weights for named
entity recognition

Word	Token position	Weight
B-PER		
$b = -4.6$		
President	−1	5.0
Clinton	0	4.5
...		
Minister	−1	4.4
...		
Prime	−2	1.4
...		
said	2	0.87
...		
B-ORG		
$b = -5.0$		
Reuters	0	5.2
U.N.	0	4.8
...		
Corp.	2	4.0
...		
joined	−1	3.1
...		
B-LOC		
$b = -5.2$		
US	0	5.7
Israel	0	5.7
...		
capital	−1	3.4
southern	−1	2.5
...		

the context. In this example, the score for B-PER can be computed as follows:
1.4(Prime@−2) + 4.4(Minister@−1) + 0.87(said@2) −4.6(b) = 2.1. We use the
notation WORD@RELPOS to indicate that the feature associated with the token
WORD at the relative token position RELPOS (with respect to the current token) is
active.

Once weight vectors are computed, the classification rule for each feature vector
x is the label t associated with the largest score $s^t = w^t \cdot x + b^t$. Although Max-
Ent works well in practice, the GIS method converges much more slowly than the
algorithm for RRM described in Chap. 3.

6.2.4 Linguistic Features and Encoding

Strong learning algorithms are very helpful, but in order to build a successful information extraction system, it is essential to use good features that can characterize the application. We shall describe such features in this section.

Although there are specific features that are problem-dependent, some local contextual features are useful for most entity extraction tasks. We shall describe a few types of features below. Consider a sequence of tokens $\ldots, tok_{-2}, tok_{-1}, tok_0, tok_1, tok_2, \ldots$. Assume that we have assigned class labels t^i for each token tok_i ($i < 0$), and we are examining the current token tok_0.

- Tokens within a window (typical window size is two tokens to the left and right of the current token): These features provide contextual linguistic information that is useful for determining the class to which entity the current token belongs.
- Previously assigned class labels (typically the previous two labels): Since certain label sequences are more likely to happen than other label sequences (and some sequences are prohibited), these features capture some information on the interdependency of the label sequence.
- Token-based annotations from existing linguistic processing modules: for example, the part of speech of the token or whether a token sequence matches a certain linguistic pattern. Any additional linguistic annotation can be regarded as a feature. However, the following specific features are relatively simple and often very useful:
  - Token type features: Semantic or syntactic type information can provide useful information; for example, whether the token is capitalized, whether the token denotes a number, or whether it is in a specific semantic class.
  - Character-based token features: These are character subsequences inside a token, such as a prefix or suffix up to a certain length. These features capture the composition of the word, which often yields useful information on its semantic class. For example, for named entity recognition, one can sometimes guess whether a word is a person name by simply looking at its character composition, even though one has not seen the word before.
  - Dictionary features: This feature determines whether a token phrase belongs to a certain prebuilt dictionary or not.
- High-order conjunction of basic features: These are occurrence features. For example, instead of taking token tok_{-1} and tok_0 as separate features, we may use their conjunction, which is essentially the token phrase $\{tok_{-1}, tok_0\}$. As another example, instead of features expressing whether a token belongs to a first-name-dictionary FD or a last-name-dictionary LD, we may create a conjunction feature that checks whether tok_0 belongs to FD and tok_1 belongs to LD.

For each data point (corresponding to the current token tok_0), the associated features are encoded as a binary vector x, which is the input to the machine-learning algorithm. Each component of x corresponds to a possible feature value v of a feature f such as described above. The value of the component corresponds to a test that has value one if the corresponding feature achieves value v or value zero if the corresponding feature achieves another feature value.

For example, since tok_0 is in our feature list, each possible word value v of tok_0 corresponds to a component of x: the component has value one if $tok_0 = v$ (the feature value represented by the component is active) and value zero otherwise. Similarly, for a second-order feature in our feature list such as $[tok_{-1}, tok_0]$, each possible value $[v_{-1}, v_0]$ in the set of all observed two-token phrases (in the training set) is represented by a component of x: the component has value one if $tok_{-1} = v_{-1}$ and $tok_0 = v_0$ (the feature value represented by the component is active), and value zero otherwise. The same encoding is applied to other features, with each possible test of "feature = feature value" corresponding to a unique component in x.

Clearly, in this representation, the high-order features are conjunction features that become active when all of their components are active. In principle, one might also consider disjunction features that become active when some of their components are active. Note that the representation above leads to a sparse but very high-dimensional vector. As we mentioned earlier, this requires the learning algorithm to have the ability to take advantage of large features and the sparse vector representation. For high-order features, the possible set of feature values can be very large. One way to alleviate a potential memory problem is to use a hash table. This approach works because, despite the high dimensionality of the feature space, most of these dimensions are empty.

Figure 6.7 illustrates how a feature vector is generated for the token *German*. In this example, a window size of five tokens, with the current token in the center, is used. Token positions are relative to the current token. Note that the features also encode positional information: "*to* at position $+1$" is a different feature from "*to* at position $+2$."

6.2.5 Local Sequence Prediction Models

Although it is possible to obtain good performance by separately predicting the label associated with each token, it is often better to take advantage of the interdependency among the label sequences, and jointly maximize the most likely label sequence. This can be achieved by sequential probability modeling and dynamic programming.

We shall briefly describe the idea without going into the technical details. Note that, in our model, the feature vector x^i can depend on the previous c class labels $t^{i-1}, \ldots, t^{i-c}$. Therefore, we can write the conditional probability for t^i as

$$\Pr(t^i | \text{sentence}) = \Pr(t^i | \tilde{x}^i, t^{i-1}, \ldots, t^{i-c}), \qquad (6.5)$$

where $\tilde{x}^i$ is the part of x^i that does not depend on the label sequence $[t^j]$, and x^i is a concatenation of $\tilde{x}^i$ and $t^{i-1}, \ldots, t^{i-c}$. For notation simplicity, in the following, we will not differentiate x^i from $\tilde{x}^i$, and simply denote the t^i independent feature vector by x^i. This should not cause any confusion.

Token Sequence: EU rejects German call to boycott British lamb .

Feature Type	Nonzero Features for *German*
Previous two labels	*Tok* − 2 is labeled I-ORG
	Tok − 1 is labeled O
Initial capitalizations in window of ±2	Current token starts with a capital letter
	Tok − 2 starts with a capital letter
All capitalizations in window of ±2	*Tok* − 2 is all capitalized
Prefix strings of length ≤ 4 of current token	G
	Ge
	Ger
	Germ
Suffix strings of length ≤ 4 of current token	rman
	man
	an
	n
Positional tokens in window of ±2	*German* at position 0
	call at position +1
	to at position +2
	rejects at position −1
	EU at position −2

Fig. 6.7 Tokens to feature vectors

Fig. 6.8 Graphical
representation of local
discriminative sequence
prediction model

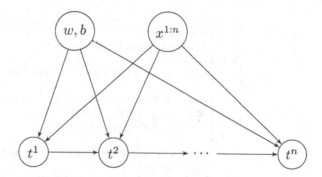

In the literature, this model is often referred to as local discriminative model, which means that each t^i depends locally on a few previous class labels (in addition to the standard global feature vector x^i which is independent of the label sequence. The graphical representation of the local dependency relationship, for $c = 1$, is in Fig. 6.8. In this figure, $x^{1:n}$ represents the information contained in the original sentence that is independent of labels $\{t^i\}$. The arrow from t^{i-1} to t^i means that each label depends only locally on its previous label.

Once we have learned a local discriminative model, we can find the optimal label sequence $t^{1:n}$ for a given sentence $x^{1:n}$. This process is called *decoding*. Instead of finding t^i one by one sequentially from $i = 0, 1, \ldots$, it is common to search for a label sequence $\{t^i\}$ that maximizes the overall probability

$$\prod_i \Pr(t^i | x^i, t^{i-1}, \ldots, t^{i-c}).$$

To obtain the exact solution, one can use dynamic programming, also referred to as the *Viterbi* algorithm. However, in practice, it is often enough to do so approximately with a simpler procedure. Instead of keeping just one sequence $\{t^i\}$ at each point, one may keep the top-k sequences $\{t^i\}$ ranked by the joint probability seen so far. When predicting the current label, we use all the top-k sequences. We then rerank the new sequence with the current label included and then only keep the top-k label sequences. We continue this process until finishing the last token. Then we output the top-ranked label sequence as the final prediction. The sequential tagging-based entity extraction is summarized in Fig. 6.9.

To see how this works, we shall illustrate the idea by a very simple example in Table 6.3. Assume that we encounter the phrase *New Mexico State University* in the named entity recognition task. For simplicity, we assume that the features we use for each token are merely the previous, the current, and the next token, as well as the label of the previous token. The rows of the table are the stages in the decoding process where we look at each token from left to right one by one. The token currently under examination is listed in the first column. The top two choices of probable labels are listed in the remaining columns, where we use the format of first choice/second choice.

```
• Tokenize the text.
• Create a feature vector for each token.
• In training:
  – Encode the entities as a label sequence aligned with to-
    kens.
  – Create (token feature, token label) training pairs.
  – Use a learning algorithm to learn a rule that can output the
    probability of token label conditioned on token feature.
• In decoding:
  – Find label sequence that approximately maximizes the
    joint conditional probability.
  – Translate labels back into extracted entities.
```

Fig. 6.9 Sequential tagging-based entity extraction

Table 6.3 Decoding example

Current token	New	Mexico	State	University
New	B-LOC/B-ORG		?	
Mexico	B-LOC/B-ORG	I-LOC/I-ORG	?	
State	B-ORG/B-LOC	I-ORG/I-LOC	I-ORG/B-ORG	?
University	B-ORG/B-LOC	I-ORG/I-LOC	I-ORG/B-ORG	I-ORG/B-ORG

When we start with the first token, *New*, the classifier can deduce based on its right context that its label is mostly likely to be B-LOC but also possibly B-ORG. If a decision is to be made immediately, the classifier will choose the incorrect label B-LOC since it is more likely. However, in the decoding stage, we will keep both choices and delay the final decision to a later stage when we can be more confident about the correct label. After we process the token *Mexico* and see the token *State*, we immediately deduce that its label should be B-ORG or I-ORG due to the right context of *University*. Since an ORG label is much more likely to follow an ORG label than an LOC label, the decoder can determine that the label sequence *B-ORG I-ORG I-ORG* for *New Mexico State* has a higher probability than *B-LOC I-LOC B-ORG*. This change is reflected in the fourth row of Table 6.3. Proceeding in this way, after seeing the last token, *University*, we are able to obtain the correct label sequence.

More generally, in a local discriminative model, one may employ non-probabilistic prediction models by reducing sequence prediction into a standard classification problem. In this approach, we simply predict the next label t^i given the previous label t^{i-1} and the observation $x^{1:n}$. One may use any classification algorithm such as SVM to solve this problem. The (linear) scoring function returned by the underlying classifier can then be used as the scoring function in the Viterbi decoding algorithm or its approximation described above.

Similar to the relationship of generative model (naive-Bayes) versus discriminative model (maximum entropy) in classification, one can also consider the generative version of the local sequence model, which has a graphical representation in Fig. 6.10 (for simplicity, we do not plot the dependency of x^i on model parameters (w, b)). Such local generative sequence models are traditionally referred to as hidden Markov models, where at each position, we model the relationship of x^i given label t^i, as well as local dependencies of t^i and t^{i-1} (or $t^{i-1}, \ldots, t^{i-c}$). Here we may still employ the naive-Bayes model for $P(x^i | t^i)$. However, in practice, the local discriminative sequence models performs better than local generative sequence models because it is easier to incorporate large number of features in x^i in discriminative models without making unrealistic independence assumptions among the features (as in naive-Bayes like generative models).

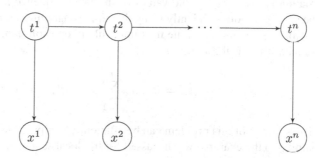

Fig. 6.10 Graphical representation of hidden Markov model

6.2.6 Global Sequence Prediction Models

The conditional probability model for the sequence $\{t^i\}$ is decomposed in (6.5) as the product of local dependencies:

$$\Pr(t^{1:n}|\text{sentence}) = \prod_{i=1}^{n} \Pr(t^i|x^i, t^{i-1}, \ldots, t^{i-c}).$$

It corresponds to the graphical model in Fig. 6.8 with $c = 1$. Another approach is to treat the prediction of label sequence $t^{1:n}$ directly as a multi-class classification problem. If each label has k values, then the sequence $t^{1:n}$ can take k^n possible values. We can then directly generalize the MaxEnt model (6.2) to this k^n-class multi-category classification problem using the following representation:

$$p(t^{1:n}|w, b, x^{1:n}) = \frac{e^{f(w, x^{1:n}, t^{1:n})}}{\sum_{t^{1:n}} e^{f(w, x^{1:n}, t^{1:n})}}, \tag{6.6}$$

where

$$f(w, x^{1:n}, t^{1:n}) = \sum_{i=1}^{n} w \cdot z_i(t^i, t^{i-1}, \ldots, t^{i-c}, x^{1:n}).$$

In this model, each feature vector $z_i(t^i, t^{i-1}, \ldots, t^{i-c}, x^{1:n})$ is similar to the feature vector x^i for predicting label i in the local discriminative model. While in (6.5), we model the local conditional probability $p(t^i|t^{i-1}, \ldots, t^{i-c})$, which is a small fragment of the label sequence $t^{1:n}$, in (6.6), we directly model the global label sequence.

The probability model in (6.6) is called *conditional random field (CRF)*. More generally, it is an example of global discriminative model, where we predict the whole label sequence $t^{1:n}$ simultaneously. The graphical model representation is given in Fig. 6.11. Unlike Fig. 6.8, the dependency between each t^i and t^{i-1} in Fig. 6.11. is undirectional. This means that we do not directly model the conditional dependency $p(t^i|t^{i-1})$, and do not normalize the conditional probability at each point i in the maximum entropy representation of the label sequence probability.

The CRF model is more difficult to train because the normalization factor in the denominator of (6.6) has to be computed in the training phase. Although the summation is over k^n possible values of the label sequence $t^{1:n}$, the computation can be arranged more efficiently using dynamic programming. In decoding, the denominator can be ignored in the maximum-likelihood solution. That is, the most likely sequence $\{\hat{t}^i\}$ is the solution of

$$\{\hat{t}^i\} = \arg\max_{t^{1:n}} \sum_{i=1}^{n} w \cdot z_i(t^i, t^{i-1}, x^{1:n}). \tag{6.7}$$

The solution of this problem can be efficiently computed using approximate search, similar to the example we discussed for the local sequence model.

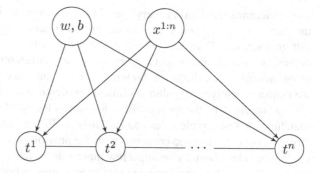

Fig. 6.11 Graphical representation of global discriminative sequence prediction model

The main advantage of global sequence model is that it is less prone to the so-called label bias problem. Although the problem is highly technical, the problem can be roughly explained as follows. In the training of a local model, we use the true labels (from previous few positions) to form each feature vector. However, in decoding, due to errors, we are not going to find correct labels for each position i, and this introduces a bias that is not present in our training data. Global models alleviate this problem by considering all label sequences (including those with errors) weighted according to their probabilities. Nevertheless, in many practical problems, the benefit of global modeling appear relatively small.

6.3 Coreference and Relationship Extraction

After entity extraction, another useful module in an information extraction system is coreference resolution and relationship extraction. Both work with the extracted entities. In coreference resolution, the purpose is to determine whether different extracted entity mentions refer to the same underlying entity. The relationship extraction module determines whether two (or more) entities have a certain prespecified relationship. Coreference resolution may also be treated as a special case of relationship extraction if we regard it as determining whether two entity mentions are identical. We give an overview of learning-based methods for these tasks. However, they are not included in the accompanying software.

6.3.1 Coreference Resolution

The main goal of coreference resolution is to group together different mentions of the same underlying entities. For example, the person *Robert L. James* may be referred to as *James* later. We want to specify that they refer to the same person and group these two entities together. A person may also be referred to by a pronoun later. For example, *Robert L. James* can be referred to as *he* later. Therefore, in coreference resolution, we would like to group them together as well.

Coreference resolution is very useful because by grouping different mentions of the same underlying entity, we can merge information obtained on this entity from different sources. Coreference can be performed either within one single document or across multiple documents. Cross-document coreference resolution is usually more difficult since the name *Robert L. James* can refer to different people. Here we confine ourselves to within-document coreference resolution.

Not surprisingly, the most useful information for coreference resolution is string matching. A few simple rules can do fairly well: we group entities with identical strings together; we group pronouns with the previous extracted entity. One can also extend the rules above by adding a few more rules, such as group a full name person entity with same last name person entities, and group a full name organization entity to its initial abbreviation organization entities, etc.

We can use a machine-learning information extraction system to improve rule-based coreference resolution. Rules such as those described above can be used as features of the learning system. The main machine-learning approach to coreference resolution is to reduce the problem into decision steps consisting of determining whether two entities belong to the same group or not. In this approach, the learning problem becomes a binary classification problem that can be handled by methods we described earlier.

If all of the binary decisions from the learning algorithm are perfect, then one can simply link every two entities that are classified to be in the same group. However, in reality, decisions from the learning algorithm are not perfect. For example, we may encounter the situation that A and B are classified to be in the same group and A and C are classified to be in the same group, but B and C are not classified to be in the same group. Such conflicts can be resolved by using heuristics, which intuitively group the entities into clusters such that pairwise connection within each cluster is strong, and pair-wise connection between the clusters is weak. Naturally, the grouping heuristics correspond to various clustering methods.

As an example, a relatively simple sequential greedy grouping algorithm works as follows. We go through entities one by one from the beginning of documents and form partial entity groups at each point. At each step, we use the learned classifier to determine the probability of the current entity belonging to the same group of every previous entity. We then compute an averaged probability score for each group and pick the group with the largest average score. If the largest score is larger than a threshold, then we assign the current entity to the group that is picked; otherwise, we create a new group with the current entity alone.

To see how the procedure described above works on a concrete example, we consider the following sentence, which contains three occurrences of organizations and one pronoun (see Fig. 6.12).

[ORG Loewen Group Inc] said on Monday that [PRONOUN it] has signed or closed acquisitions totaling US $325 million since rival [ORG Service Corp International] launched a hostile bid for [ORG Loewen] earlier this year.

We want to group the four entity mentions in Fig. 6.12 into clusters. In the pair-wise approach, we first compute scores for each pair of these four mentions as listed

Named Entity Mentions:

M1	M2	M3	M4
Loewen Group Inc	it	ORG Service Corp International	Loewen

	M1	M2	M3	M4
M1	-	0.7	−1.0	0.9
M2	0.7	-	−0.2	−0.3
M3	−1.0	−0.2	-	−1.0
M4	0.9	−0.3	−1.0	-

Fig. 6.12 Coreference named entity mentions pairwise scores

in Fig. 6.12. The scores can be obtained using a binary classifier such as the RRM method or logistic regression. A positive score means the pair is likely to belong to the same group, a score close to zero means not sure, and a negative score means the pair is likely to belong to different groups. The score of a pair that involves a pronoun is usually more difficult to estimate since this usually requires information from the sentence structure, instead of string matching alone. This difficulty is reflected in Fig. 6.12, where the pairwise scores $s(M2, M3)$ and $s(M2, M4)$ are relatively low and uncertain.

Given the scores in Fig. 6.12, we can invoke the greedy clustering scheme with a threshold of, say, zero. We start with $M1$ as a single cluster $E1$. In the next step, we examine $M2$. Since $s(M1, M2) > 0$, we put $M2$ into the cluster $E1$. We then examine $M3$. Since it has a negative score both with $M1$ and with $M2$, we determine that it does not belong to $E1$. We thus create a new cluster $E2$ that contains $M3$ only. Finally, we examine $M4$. We know it does not belong to $E2$ since $s(M3, M4) < 0$. Although it is not sure whether $M2$ and $M4$ should belong to the same cluster (and $s(M2, M3)$ takes a default value of no), there is strong evidence that $M1$ and $M4$ belong to the same cluster since they share the same partial string *Loewen*. Therefore, the averaged score is greater than zero, and we put $M4$ into the cluster $E1$.

The most important feature for determining the score of a pairwise connection in coreference resolution is string matching (and partial string matching). As we have mentioned above, if we have manually developed rules for coreference resolution, then they can also be used as features. A number of other features are also very useful, such as part-of-speech information, the syntactic head of the entity, the distance between entities, and entity type information (people, organization, location, etc.).

6.3.2 Relationship Extraction

Although relationships can exist among entities that are not in the same sentence, those long-range relationships usually require semantic inference. The relation extraction module in a typical information extraction system often is constrained to relationships among entities in a single sentence. Such relational information can be directly obtained from the syntactic structure of the sentence.

The most interesting relationships are often binary relations. For example, consider the first sentence in Fig. 6.1, where we have a relationship that the person *Robert L. James* belongs to organization *McCann-Erickson*. A binary relationship takes two typed entities as arguments. In the example above, we have a belong-to relationship that takes its first argument as a person and its second argument as an organization. In this case, the order of the arguments is not interchangeable. Some binary relationships may also take interchangeable arguments.

Learning binary relations among entities within a sentence can be posed as a binary classification problem. For each pair of entities in a sentence, we determine whether they have a certain predefined relationship or not. In many cases, one may manually write rules to match certain predefined parse structures for specific relationships. Such rules can also be used as features for a machine-learning system. Predictive methods for relationship extraction can also take other features. A very important feature is the syntactic parse tree of the sentence. One may decompose a tree into subtrees that can be encoded as features. Each of these features should also contain its relative positional information with respect to the entities.

Consider the following sentence, with named entities extracted. There are three entities, and we are trying to determine binary relations between each pair of them. In this example, we will focus on relations between people and thus ignore the DATE entity.

[PER Jon Smith] met [PER Mary Jones] on [DATE Sept. 17].

Assume that we obtain the following parse of the sentence, where each word is supplemented with its normalization.

[SUB [PER Jon/jon Smith/smith/]   ]
[VG met/meet]
[OBJ [PER Mary/mary Jones/jones]
      [PP on/on [DATE Sept./Sept. 17/integer]   ]   ].

We can extract the following pattern from the tree, where we use the dash symbol— to denote words that can be skipped.

[SUB – PERSON1 –] [VG – –/meet –][OBJ – PERSON2 –].

This simplified tree pattern can be used as a feature for the learning algorithm, which will then learn from the training data that a pattern of this type implies a relationship of *meet* between PERSON1 and PERSON2.

6.4 Template Filling and Database Construction

In general, a template is a more complicated structure than a binary relationship. A binary relationship contains two typed entity slots to fill, while a template may contain more than two slots, which can be organized into a semantically meaningful group. In addition, in order to fill slots in a template, we may need to go through the entire document rather than only a single sentence as in the case of within-sentence binary relation extraction.

An example of a template is an *event*, which specifies a certain occurrence associated with entities at a specific time and place. For example, a person meeting another person can be considered a *meet* event. In this case, the template has the following slots: people involved in the meeting, the location of the meeting, and the time of the meeting. This structure is more complex than one that regards *meet* as a relation between two persons, as in the example we used in the previous section. From that example, we can also note that the location of the meeting is not specified in the sentence.

Therefore, for template filling, some slots can be missing since it is possible that they are not specified in the text. In addition, slots in a certain template can be specified in different sentences. In such a case, we need to identify whether two sentences refer to the same template and then fill the template accordingly. This problem is similar to named entity coreference resolution but is often more complicated.

The traditional information extraction literature mainly focuses on problems where the whole article deals with only a single template. Often a slot of the template can be identified with a particular entity type. For example, consider extracting information from seminar announcements, where we want to know the speaker, talk title, time, and location. The slots are extracted from an article, and each can be associated with a different entity type. Therefore, once we extract the entity types, the template can be easily filled using a simple rule that fills each slot with the corresponding entity extracted from the text. Consider the following short seminar announcement.

Dr. Jon Smith from the IBM text analysis group will present a seminar entitled "The future of predictive text-mining" on Tuesday, Sept. 2nd from 11:00 am–12:00 am in the auditorium.

We may create entity types SPEAKER (the speaker), ORGANIZATION (where the speaker is from), TITLE (the title of the talk), DATE and TIME (date and time of the talk), and LOCATION (the place of the talk). The announcement can then be tokenized and annotated using a named entity recognizer as:

Dr. [SPEAKER Jon Smith] from the [ORGANIZATION IBM text analysis group] will present a seminar entitled " [TITLE The future of predictive text-mining] " on [DATE Tuesday, Sept. 2] from [TIME 11:00 am–12:00 am] in the [LOCATION auditorium].

The desired template can be easily filled from the annotation above.

6.5 Applications

6.5.1 Information Retrieval

One interesting application of information extraction is automated citation analysis for academic papers. A notable example is the Citeseer search engine. The underlying idea is to crawl academic sites and gather papers in a specific field. Then, by using information extraction technology, one can extract information such as a paper's title, author, abstract, and bibliography list, as well as the context in which citations are made. Further statistical and linkage analysis can then be applied on the extracted information and provided as feedback to the users.

Another application of entity detection is natural language question answering. Assume we ask the question: *How far is it from Los Angeles to San Francisco?* The system answers *382 miles*, based on the following sentence from its database: *The distance between Los Angeles and San Francisco is 382 miles.* A question-answering system classifies questions into various types and preprocesses its database to extract entities marked with appropriately defined answer types that correspond to different question types. In the example above, *Los Angeles* and *San Francisco* will be marked as entities of city type, and *382 miles* will be marked as an entity of distance (or measure) type. From the question, the system first determines that the answer should be a distance between two places called *Los Angeles* and *San Francisco*. These places are then matched with sentences in the database, and then the entity of distance type *382 miles* is extracted as the answer.

6.5.2 *Commercial Extraction Systems*

A number of small companies each have a fraction of the market for off-the-shelf named entity extraction systems. Let us briefly look at three of them.

IdentiFinder is an incarnation of longstanding work in the area by Bolt, Beranek and Newman (BBN). IdentiFinder uses a Hidden Markov model to recognize entities. Each word in a text is labeled with one of the desired class names or with Not-A-Name. Separate bigram probabilities are generated for each type of entity. Statistical models of this kind require substantial training data.

NetOwl is an outgrowth of work by SRA, Inc., on a system originally called NameTag. SRA founded IsoQuest to market NetOwl. The NetOwl extractor goes beyond the extraction of names to events connecting people and organizations to items such as weapons of mass destruction. A particular instance of NetOwl is created by writing a set of declarative data files that contain the task-specific content. These are compiled into a configuration file that is read by the extraction engine. The extraction engine, in consequence, is completely application-independent. Actual name identification is accomplished by hand-crafted rules and dictionaries.

ClearForest is an Israeli company that has made a number of sales to American corporations and government agencies. Its software scans text and tags items of interest for display to the customer. The tagging portion uses semantic, statistical, and structural clues to classify documents and extract entities and relationships. An application called *ClearResearch* organizes and presents the extracted data. A recent module, *Link Analysis*, finds indirect links between entities. The extraction module is based on "rulebooks," handwritten rules specific to an application or customer. A set of prewritten rulebooks is available for life sciences, counterintelligence, intellectual property, and financial services.

6.5.3 Criminal Justice

Two prototype applications in the area of criminal justice have been described in the open literature. One was sponsored by the US Department of Justice's National Law Enforcement and Corrections Technology Center. The project involved scanning documents seized in raids and converting them to computer text files using optical character recognition. The structured information (people, places, organizations) from these papers is then entered into a database where standard visualization and analysis tools can be used. The project was designed to investigate named entity extraction for structuring. It used the commercial package IdentiFinder to extract entities in a money-laundering case. The final report included a list of what would be needed to put such a system into production.

In a similar project, the University of Arizona tested its neural-network-based extraction system on police narrative reports from the Phoenix Police Department dealing with narcotics crime cases. In this case, the entities were person, address, vehicle, drug, and personal property. The named entity tagger did well on person (precision 0.741, recall 0.739) and drug (precision 0.854, recall 0.779) and less well on address (precision 0.596, recall 0.574) and personal property (precision 0.488, recall 0.478). It was hypothesized that performance on addresses could be greatly improved, but identifying personal property entities was inherently hard.

On the boundary between military and criminal applications is a project called the Automated Counterdrug Database Update (ACDBU) program. It was established by the Joint InterAgency Task Force on drug interdiction. The goal is to streamline and automate counterdrug and maritime tracking operations. The program is capable of extracting data from all military intelligence formats as well as from ordinary text. Example entities extracted are names, locations, dates, and money values. The underlying software is IdentiFinder. Recognition rules are automatically trained rather than being handwritten finite-state rules.

6.5.4 Intelligence

Application of named entity tagging and information extraction for problems in military intelligence and homeland security should not be surprising since much of this research has been funded by the Defense Department. One prominent contractor to the military and to Homeland Security is Inxight Software Inc., a spinoff from Xerox

Corporation to capitalize on 20 years of linguistic and artificial intelligence work by its Palo Alto Research Center. According to a February 2003 press release, "using Inxight software, intelligence analysts might locate previously unknown people associating with known terrorists/criminals or known terrorist/criminal organizations." The software is used to process hundreds of messages per day (intelligence reports, e-mails, news articles, Web pages) to look for these links. In some cases, a simple listing of the entities occurring in a document allows analysts to find documents of interest simply by looking at the list rather than reading the whole document. The software supports multiple languages, including Arabic.

This kind of software is deemed so important that the CIA has established a venture capital firm (In-Q-Tel Technologies) to support small companies working in the area. One example is a company called Attensity, whose software extracts common threads of information out of documents. Another example is Systems Research and Development (SRD), which specializes in link analysis using a software package called NORA (Non-Obvious Relationship Awareness) that traces links across data banks between entities such as people, places, and organizations. This software was originally developed for use by Las Vegas casinos to spot cheaters and card counters by tracing links between individuals and earlier transactions. For example, it might tell a casino that a job applicant once had the same address as a known criminal.

Although a detailed description of the application is not available, the FBI has purchased ClearForest products in its counterterrorism data system, comprising over a billion documents, with up to 1000 new documents each month. It is available on the desktops of all its analysts, delivering visual, interactive summaries.

Named entity extractors can certainly have uses other than in intelligence and criminal justice. One prototype system is designed to extract entities from Thai agricultural documents to improve information retrieval. Another project, in Europe, is intended to extract information in English, Greek, French, and Italian from Web pages to facilitate product comparisons. A third example is formatting the information in classified ads. Electronic Classifieds, Inc., used NetOwl as part of a system for formatted access to the information in ads. For a car ad, NetOwl extracted year, make, model, transmission, color, mileage, price, etc. Knight Ridder newspapers uses a similar system for putting ad content into an object database, from which it could more easily be retrieved. In the Knight Ridder system, there are up to 70 attributes for a vehicle. One of the problems at Knight Ridder is that their 31 papers use a variety of front-end systems. This means that the same content can have different descriptions (e.g., "bedroom," "bdrm," "bdr," "bed"). NetOwl is able to handle these varying descriptions.

6.6 Summary

A common task in natural language processing and text mining is the extraction and formatting of information from unstructured text. One can think of the end goal of information extraction in terms of filling templates codifying the extracted information. The templates are then put into a knowledge database for future use. This chapter describes several models and learning methods that can be used to solve information extraction. We focused on two major subtasks, one is to extract entities, such as person name, organization, etc. from sentences, and the other is to determine the relationship among extracted entities.

6.7 Historical and Bibliographical Remarks

The field of text-based information extraction has a long history in the artificial intelligence and computational linguistics areas. However, modern evaluations and progress of information extraction have greatly benefited from the DARPA-sponsored Message Understanding Conferences (MUC) from the late 1980s to late 1990s. The last MUC was MUC-7. NIST still maintains a Web site for MUCs at http://www-nlpir.nist.gov/related_projects/muc/index.html that includes the proceedings of MUC-7. This final evaluation includes results for IdentiFinder and NetOwl. In the MUC-7 evaluation (Krupka and Hausman 1998; Miller *et al.* 1998), IdentiFinder was trained on 790,000 words of New York Times newswire data. In the evaluation, IdentiFinder had an F-measure of 90.44. On the same evaluation, NetOwl achieved an F-measure of 91.6. The high F-measure for NetOwl and IdentiFinder is the result of careful tuning on a well-understood task.

Multiple universities and organizations have participated in conferences such as MUC, which helped to form the field of information extraction. After MUC, NIST sponsored a new ACE program, which stands for *automatic content extraction*. Although similar to MUC, the task definitions are different. More information on ACE can be found at http://www.itl.nist.gov/iad/894.01/tests/ace/.

Many of the earlier systems relied heavily on rule-based extraction patterns. See, for example, AutoSlog (Riloff 1993), LIEP (Huffman 1995), PALKA (Kim and Moldovan 1995), CRYSTAL (Soderland *et al.* 1995), RAPIER (Califf and Mooney 1998), SRV (Freitag 1998), and WHISK (Soderland 1999). Such systems often directly target at template filling.

One disadvantage of this approach is that extraction rules are usually relatively simple. It is quite difficult to take advantage of large features that incorporate different information sources. Emphasis has shifted to more complicated machine-learning methods as well as combining machine-learning methods with rule-based systems. For named entity extraction, machine learning methods were used in several systems in the later MUC conferences. For example, BBN's IdentiFinder was based on HMM (Bikel *et al.* 1999), and CMU's MENE system was based on maximum entropy (Borthwick 1999). In MUC-7 named entity extraction tasks, the best performance was achieved by using a combination of manually developed rules and MaxEnt learning (Mikheev *et al.* 1998). Some more recent evaluations on multilingual named entity extraction have been organized in the form of shared tasks as part of the Conference on Natural Language Learning (CoNLL). The data sets used in the shared tasks are publicly available from http://www.cnts.ua.ac.be/conll2003/ner/.

Coreference resolution was part of later MUC and the new ACE tasks. A number of learning-based approaches appeared in the literature. One of the earliest machine-learning systems was based on decision trees (McCarthy and Lehnert 1995) (also see Soon *et al.* 2001). The clustering view of coreference resolution was explicitly discussed in Cardie and Wagstaff (1999). This problem is also closely related to Anaphora Resolution (Lappin and Leass 1994), which has been widely studied in the computational linguistics literature. Relational learning can be regarded as a special case of template filling. Systems for learning extraction patterns, such as RAPIER, mentioned above, can be used for this purpose. See Craven and Slattery (2001), Roth and Yih (2001), Zelenko *et al.* (2003) for additional automated approaches to relational learning.

The principle of maximum entropy has a long history (Jaynes 1957). Non-probabilistic discriminative local sequence models have also been widely used. An example is Zhang *et al.* (2002). The global model CRF was introduced in Lafferty *et al.* (2001). McCallum's Mallet software package, obtainable from http://mallet.cs.umass.edu, contains an implementation of CRF. More generally, global discriminative learning refers to the idea of treating sequence prediction as a multi-category classification problem with k^n classes, and a classification rule of the form (6.7). This approach can be used with some other learning algorithms such as Perceptron (Collins 2002) and large margin classifiers (Taskar *et al.* 2004; Tsochantaridis *et al.* 2005; Tillmann and Zhang 2008). The GIS algorithm that we presented was proposed and analyzed in Darroch and Ratcliff (1972). Although new methods such as improved iterative scaling (Pietra *et al.* 1997) have been proposed, GIS is still widely used in practice. Risk minimization and large margin methods have become popular due to the success of support vector machines (Vapnik 1998) and boosting (Freund and Schapire 1997; Breiman 1999; Hastie *et al.* 2001). They can be regarded as generalizations of MaxEnt and can be applied to information extraction problems. For example, a version of robust risk minimization was applied to the text-chunking problem (Zhang *et al.* 2002). The algorithm described in Chap. 3 was used in the CoNLL 2003 named entity shared task (Florian *et al.* 2003; Zhang and Johnson 2003), where results reported in Florian *et al.* (2003) achieved the top performance among the 15 participating systems.

Details on commercial systems are rather sketchy, being closely held trade secrets. In some cases, the basic ideas are known because the systems have their roots in publicized university or research lab work, but one never knows how much reworking has been done. Further details can be gleaned from the Web pages of the companies mentioned and from brief news items in the computer trade press.

6.8 Questions and Exercises

1. Give any two language-dependent features that might be appropriate for a system that extracts named entities from documents.
2. Which of the software tools would need to be replaced in order to create a feature vector for named entity extraction?

3. Run the example NE extractor on some test data (you can use any text file for this purpose). Why are the results not better than they are?
4. Using the advanced search module of your preferred search engine, look at the first 50 results for the search "* KFC * China *". How many possibly useful extraction patterns were found?
5. Starting with a potentially relevant and specific pattern, like "* set to * open in *" or "* restaurants in *", see how many out of 50 results are relevant at all and how many restaurant pairs are found.
6. Some patterns are quite productive. Try "* merger of *" and note the number out of 50 responses which refer to companies. Still, there are a significant number of responses to filter.

Chapter 7
Data Sources for Prediction: Databases, Hybrid Data and the Web

Our explication of predictive text mining has focused on data sources that are collections of documents. It's a pure world where all data are texts composed of words and alphanumeric characters, not numbers. Similarly, the world of classical statistics and data mining expects purity of data, composed of spreadsheets, tables and numbers, where text is an unnatural fit. Data mining is characterized as processing structured data, and text mining works with unstructured data. In this chapter, we relax this separation and consider different types of mixed data, both numerical and text.

Collecting data and preparing for analysis can be tedious and complex tasks. Predictive text mining is often high-dimensional, making a carefully assembled and large sample critically important. The web is an increasingly valuable source of documents. For some web-based data sources, documents arc formatted in a standardized form, readily amenable to further processing and mining. In this chapter, we discuss the adaptation of predictive mining techniques to web-based data, amalgamating structured, hybrid and unstructured data.

7.1 Ideal Models of Data

7.1.1 Ideal Data for Prediction

Most machine learning methods for prediction expect data in the form of a single table, such as illustrated in Fig. 1.2. This is equivalent to the ubiquitous spreadsheet in csv (comma-separated values) format that represents a sample of data. Each row is an example, like a patient record, and each column is a recorded measurement or finding. All entries are ordered numerical variable or categorical, true-or-false variables. One column is reserved for the target or label and is either true or false for classification, or ordered numerical for regression.

The prediction model anticipates structured data, and the learning methods are highly developed. Data sources abound. For smaller applications, spreadsheets may

S.M. Weiss et al., *Fundamentals of Predictive Text Mining,*
Texts in Computer Science 41,
DOI 10.1007/978-1-84996-226-1_7, © Springer-Verlag London Limited 2010

Table 7.1 Useful Linux commands

cut	cut out the specified columns from a table
paste	paste the rows of two tables together
join	join two tables using a key
sort	sort rows of table

Table 7.2 Transforming XML to structured table

Produce dictionary = feature set

Measure numerical property of each feature = column

Record values for each document = rows

contain the full sample. For larger problems, databases with linked tables are a ready source of structured data. Some additional data preparation is typically needed, specialized to the application. Linux commands, such as those listed in Table 7.1 are highly effective in assembling the single table needed by prediction methods. A good portion of many data-mining efforts is spent on transforming data into this ideal format. When data reside in a database or spreadsheet, the conversion is not usually difficult.

7.1.2 Ideal Data for Text and Unstructured Data

For text, the ideal representation is in XML, such as the example illustrated in Fig. 2.1. Within each document, the label is identified, and the body of alphanumeric text is presented. This hardly looks like the tables described previously for structured data that are ideal input for predictive models. Yet, the end result is the same. The XML data are translated into a table that fulfills a tabular representation for structured data. Once text is wrapped in an XML format that is consistent for each document, it could be considered structured data, mappable into a table. Table 7.2 summaries the steps of the transformation of XML to a numerical format. Words or tokens are features and they are measured in the tabular columns. A measurement can be as simple as the presence (1) or absence (0) of the word in a document. Measurements are made for each document and recorded in a row or table.

7.1.3 Hybrid and Mixed Data

In many instances, numerical and text data can be merged into a shared format for prediction. In the simplest situation, tables derived from alternative numerical or textual sources have the same number of rows, and the tables can be pasted together. While it is natural to consider analysis of numerical and text data as distinct and

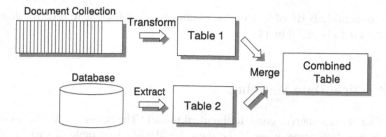

Fig. 7.1 Merging data from documents and databases

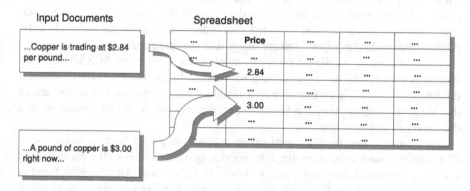

Fig. 7.2 Extracting values from documents to fields in tables

independent tasks, a hybrid approach that combines both sources can be beneficial and enhance predictive quality.

The hybrid approach combines tables from difference sources, all related to the same topic and prediction label. It does not matter whether the original data originate in database tables or documents. They all complete their voyage as simple tables with measurements on the same examples. They can be combined as illustrated in Fig. 7.1.

An alternative view of text mining is to populate a database by browsing through documents and extracting the information necessary to populate a field or measurement in the table. We have referred previously to this task as "information extraction." For instance, we may be composing a table where each row is an example of a product. We might roam the web to compose this table. One important feature might be the current product price. This mission may imply knowledge of how to find the price in the document and then to extract it and enter in the table. Some data may be readily available from sources in tabular formats. Others may be dynamically extracted from web pages. Figure 7.2 illustrates extracting a single value from documents and then populating a field in a table. The original data sources are mixed collections of structured and unstructured documents. This scenario can be more complex than accessing a database because specialized knowledge is usually required for information extraction. A simple XML representation of just a label and the remaining body of text in documents is not sufficient to directly describe fields that are to be filled in a table.

7.2 Practical Data Sourcing

How close do data sources come to the ideal model? The better the fit, the less need for extensive data preparation. If the data are already in a table, a variant of csv format, then the path to an ideal format is usually clear. That does not mean that no additional transformations will be necessary. Rather, these transformations will be perceived as relatively close to an ideal data format for prediction. Because much structured data, i.e. numerical or coded data, resides in databases or spreadsheets, predictive modeling is a natural fit for many real-world database applications.

When documents in digital form are labeled, the conversion to XML for prediction is trivial. Tags are confined to a few simple fields like the body of text and the label. Figure 7.3 is an example of a standard representation. Once data are in this simplified format, they can be translated directly into a tabular format using a dictionary of words as features.

The natural home for structured numerical data is a data base or spreadsheet. The source of such data, often through proprietary databases, is well-established. In contrast, sources of documents are increasingly web-based. An emerging trend is the greater use of richer XML notation, where many tags are specified, giving much greater structure to document content. This additional structure facilitates information extraction. For example in Fig. 7.3, which is an excerpt from a research paper, has a tag for the author names.

Extracting documents manually from the web is obviously too cumbersome to be practical. Different mechanisms can be found to automatically download a web page and related pages. For illustration, we will look at a built-in Linux command, *wget*, that performs this function:

```
wget [options] URL
```

The wget command has many options. Try searching for them on the web. The simplest form just downloads the page at the specified URL. Using wget, a script could be composed to download automatically specific pages en masse from the web. If these pages all contain a field in a consistent location, preceded and followed by certain tags or keywords, then a Linux command like *awk* could be used for information extraction. The following command extracts all text between *string1* and *string2*:

```
awk -/string1/,/string/
```

We have seen that numerical or coded data are readily obtained from databases and are consistent with a data model for prediction. Documents can be formatted as a simplified XML specification and then mapped into a standard table that is also consistent with prediction. If matched properly, data from both of these sources can be amalgamated into a hybrid model of data. The web is a massive and growing

```
<DOC>
<TITLE>
Exercise therapy in Type 2 diabetes
</TITLE>
<AUTHOR>
Praet SF, van Loon LJ.
</AUTHOR>
<SUBJECTS>
<SUBJECT>Pathophysiology</SUBJECT>
<SUBJECT>Type 2 diabetes</SUBJECT>
</SUBJECTS>
<TEXT>
Structured exercise is considered an important cornerstone to
achieve good glycemic control and improve cardiovascular
risk profile in Type 2 diabetes. Current clinical guidelines
acknowledge the therapeutic strength of exercise intervention.
This paper reviews the wide pathophysiological problems
associated with Type 2 diabetes and discusses the benefits of
exercise therapy on phenotype characteristics, glycemic
control and cardiovascular risk profile in Type 2 diabetes
patients. Based on the currently available literature, it is
concluded that Type 2 diabetes patients should be stimulated
to participate in specifically designed exercise intervention
programs. More attention should be paid to cardiovascular
and musculoskeletal deconditioning as well as motivational
factors to improve long-term treatment adherence and clinical
efficacy. More clinical research is warranted to establish the
efficacy of exercise intervention in a more differentiated
approach for Type 2 diabetes subpopulations within different
stages of the disease and various levels of co-morbidity.
</TEXT>
</DOC>
```

Fig. 7.3 PubMed XML example. With kind permission from Springer Science+Business Media: *Acta Diabetologica*, Exercise therapy in Type 2 diabetes, vol. 46, pp. 263–278, 2009, Stephan F.E. Praet and Luc J.C. van Loon

source of documents that may be used advantageously in text mining, especially when increasing numbers of sites are structuring text in richer forms of XML. In the next section, we examine a few scenarios to illustrate these concepts.

7.3 Prototypical Examples

The web is a rich source of non-proprietary information, some of which is close to our desired representation for data. Data that was once expected to be unstructured, when accessed through the web is seemingly structured or readily translated into a tabular format. We will look at several examples of these concepts in action. They are not meant to be complete. Rather, they illustrate the potential for preparing data found on the web that may be combined with data from alternative and structured

Table 7.3 Google's stock price data from Yahoo! Finance	Date,Open,High,Low,Close,Volume,Adj Close
	2009-10-28,547.87,550.00,538.25,540.30,2567800,540.30
	2009-10-27,550.97,554.56,544.16,548.29,3216500,548.29
	2009-10-26,555.75,561.64,550.89,554.21,2970400,554.21
	2009-10-23,555.25,557.89,551.20,553.69,2392700,553.69
	2009-10-22,550.00,555.00,548.00,554.09,2336500,554.09
	2009-10-21,549.91,559.35,549.00,551.10,3670600,551.10
	2009-10-20,551.64,552.95,540.70,551.72,4043700,551.72
	2009-10-19,552.69,553.60,548.73,552.09,3217900,552.09
	2009-10-16,547.33,554.75,544.53,549.85,8841900,549.85
	...

sources. In the examples that follow, the format is correct at the time of writing. A web site could change their format, requiring some modifications to reproduce the same results. For each example, we provide an auxiliary Linux script that implements the procedures that download the data or output results.

7.3.1 Web-based Spreadsheet Data

Many sites have downloadable data in csv format. This is common for financial data. Using the following command, typed on a single line, the complete history of Google's stock price can be downloaded from yahoo.com:

```
wget -O goog.csv http://ichart.finance.yahoo.com/table.txt?s=goog&
a=00&b=2&c=1987&d=12&e=2&f=2009&g=w
```

The required input format can be obtained by manually going to yahoo. The data are downloaded to goog.csv in the format of Table 7.3.

These data are close to the standard prediction format. The are inherently numerical, not text. For prediction, adjustments and transformations will be made depending on application goals. For example, a goal might be to predict the price one month from now. The features used might be an average of prior values within a 2 week window. In this discussion, our purpose is not to describe a specific prediction problem. Rather, the objective is to obtain data and transform that data into a format suitable for prediction. If the data can be saved in csv format, then they are likely suitable for machine-based prediction with some massaging.

7.3.2 Web-based XML Data

Obtaining numerical data in standard csv format is no surprise. Spreadsheets and tables are ubiquitous, and it is natural for web sites to prepare data in this style. Sim-

```
# usage: getpubmed searchterm
# example: getpubmed h1n1 >h1n1.xml
# set the environmental variable http_proxy appropriately if using
# a web-proxy.
#
wget -O- "http://eutils.ncbi.nlm.nih.gov/entrez/eutils/
esearch.fcgi?db=pubmed&term=$1" |
grep "<Id>" |
sed 's/<Id>\(.*\)<\/Id>/http:\/\/eutils.ncbi.nlm.nih.gov\/entrez\/
eutils\/efetch.fcgi?db=pubmed\&id=\1\&retmode=xml/' >temp
wget -i- -O- <temp |
awk -v tag=$1 'BEGIN{printf("<%s>\n",tag)} !/ArticleSet|\?xml/
{print $0} END{printf("</%s>\n",tag)}'
```

Fig. 7.4 The script *getpubmed*

ilarly, a structured description of text can be posed in XML and these data can be downloaded. Once text is described in XML, and the style is consistently recorded for all documents, the complexity of addition data prep is not great. Just as some web sites provide numerical data in csv format, some provide text in XML format. A natural candidate for dispensing a rich description of documents is a literature repository. Here is an example from the MEDLINE research paper library, developed by the National Library Of Medicine of NIH.

The NLM maintains an online library, PubMed, of research articles related to medicine and life sciences at medlineplus.gov. A related URL of interest for researchers is http://eutils.ncbi.nlm.nih.gov. A query can be submitted and article abstracts can be retrieved from this library. Implicitly using wget, a query might be submitted by executing the script *getpubmed* of Fig. 7.4 as shown below, where the search term is H1N1:

```
getpubmed h1n1
```

The related articles are retrieved using the appropriate options, and the document format is XML. Figure 7.5, is an abstracted example of one the documents for h1n1. Because these documents are nicely formatted, they are amenable to information extraction. For example, the title of each document is tagged and can be extracted for each document. In Fig. 7.6, the words in the title of each document are submitted to a Google search and the hit count is recorded. The goal is to find a rough measure of the potential uniqueness or relevance of each of the articles on H1N1—are they new concepts, well-studied areas, or topics of special interest?

The PubMed data are exceptionally well-organized for both prediction and information extraction. It is a preferred representation of text and documents, especially with a uniform tagging scheme. While an emerging trend in document storage is to provide enhanced XML descriptions, most documents will be encountered in straightforward text format that is not captured clearly by XML tags and descriptors.

```
<PubmedArticle>
    <MedlineCitation Owner="NLM" Status="Publisher">
        <PMID>19693661</PMID>
        ...
        <Article PubModel="Print-Electronic">
            ...
            <ArticleTitle>Molecular evolution of novel swine-origin
            A/H1N1 influenza viruses among and before
            human.</ArticleTitle>
            ...
            <Abstract>
                <AbstractText>We find that the novel A/H1N1
                influenza viruses exhibit very low genetic
                divergence and suffer strong purifying selection
                among human population and confirm that they
                originated from the reassortment of previous
                ...
                the early evolution of this virus.</AbstractText>
            </Abstract>
            ...
        </Article>
        ...
    </MedlineCitation>
    ...
</PubmedArticle>
```

Fig. 7.5 An abstracted H1N1 document from PubMed. With kind permission from Springer Science+Business Media: *Virus Genes*, Molecular evolution of novel swine-origin A/H1N1 influenza viruses among and before human, vol. 39, pp. 293–300, 2009, Na Ding, Nana Wu, Qinggang Xu, Keping Chen and Chiyu Zhang

1640	Clinical characteristics of paediatric H1N1 admissions in Birmingham, UK.
33600	Swine-origin influenza virus H1N1, seasonal influenza virus, and critical illness in children.
2100	Person-to-person transmission of oseltamivir-resistant influenza A/H1N1 viruses in two households; Germany 2007/08.
4000	Controlling the novel A (H1N1) influenza virus: don't touch your face!
841	Influenza epidemiology and characterization of influenza viruses in patients seeking treatment for acute fever in Cambodia.
888	Oseltamivir-Resistant Novel Influenza A (H1N1) Virus Infection in Two Immunosuppressed Patients—Seattle, Washington, 2009.

Fig. 7.6 Google hit counts for titles of H1N1 documents

7.3.3 Opinion Data and Sentiment Analysis

7.3.3.1 Product Reviews

The web is a home for diverse opinions. One form of opinion that has credibility is a product review. Web sites like amazon.com encourage users to provide reviews.

```
<review>
<txt>
I bought this one a couple years ago, but not really use it until
recently. I connected it to the 12v power outlet of my car, and it
quickly blew the fuse of both of my cars. I now use a separate
12v battery to power it. It also misses a quick connect air hose,
a lot of air just leak out when I unscrew it from  the valve. I am
looking for another air inflator.
</txt>
<rating>1</rating>
</review>
```

Fig. 7.7 An example product review on amazon.com

Fig. 7.8 A dictionary extract
from negative reviews

```
blew
out
return
stalls
fuses
hose
time
tire
job
```

As an example, we examine the reviews for an air compressor used to inflate automobile tires. The compressor, used by one of the authors of this book, generally worked well, but occasionally blew the car's fuse. Amazon had 35 reader-entered reviews. No special XML here. Just a need to process the text of 35 documents and assign a label for each one. Readers rate the product on a scale of 1 to 5. We mapped these into 2 labeled groups: positive (4 or 5) or negative (1 or 2). Only 2 reviews were rated 3 and were ignored. These documents are readily downloaded from amazon.com using the following command (on a single line) repeatedly using incremental page numbers:

```
wget "http://www.amazon.com/Industries-HV40-SuperFlow-Portable-
Compressor/product-reviews/B000BMBTA2/ref=cm_cr_pr_link_next_2?
ie=UTF8&showViewpoints=0&pageNumber=2&
sortBy=bySubmissionDateDescending"
```

Figure 7.7 illustrates an excerpt of the sample following a few simple data prep steps. Thus, we have a standard prediction problem. Even before worrying about prediction, a dictionary must be assembled. Using the most frequent words (minus the stopwords) in the negatively reviewed documents, the dictionary of Fig. 7.8 is determined. Note that "fuse" and "blew" are present in this small dictionary, which might already be helpful to potential purchasers. The word "fuse" occurs in 13 of the 35 reviews, and the average rating is 3.4. In terms of prediction, we tried exploratory rule induction. While the sample is too small for reliable empirical validation, a few induced rules are described in Fig. 7.9 that discriminate negative from positive reviews.

Fig. 7.9 Decision rules for
review classification

blew & out → negative
return → negative
stalls → negative
fuses & blew → negative
OTHERWISE → positive

Fig. 7.10 Computing a
favorability index

1. For each document, compute $S = p/(p + n)$, where p is the sum of all positive term frequencies and n is sum for negative term frequencies.
2. Classify as positive document if $S > t$, some threshold, usually 50%. Otherwise, classify it as a negative document.
3. Compute the favorability index, FI based on global ratio of all S. $FI = $ (number of positive documents)/(number of negative documents).

7.3.3.2 Sentiment Analysis

A related, but more general type of opinion evaluation is known as *sentiment analysis*. Here the objective is to peruse data on the web and determine whether sentiment is favorable or not. Just like standard polling, the targets can be wide-ranging. For example, a favorability index could be computed to measure a company's overall brand image, or a product's image or perhaps public opinion of a political issue. Unlike regular polling, no individuals are interviewed. Instead, text and documents from the web are reviewed. The data sources could be blogs for political analysis, professional product reviews or discussion groups for product and company image analysis. The possibilities for sentiment are unlimited, but the conclusions are often restricted to a favorability measure.

The analysis of reviews described in Sect. 7.3.3.1 could be considered a form of sentiment analysis. Those reviews were labeled, thereby driving our understanding of positive and negative words and phrases. For many applications of sentiment analysis, massive amounts of data may be processed, and these documents are unlabeled. A favorability index can still be computed when a dictionary is composed using prior knowledge or experience. For example, two dictionaries are compiled, one with favorable words and a second one with unfavorable words. Very general dictionaries of these types have been composed independent of application. Clearly, they are not as specific as those obtained from specialized product reviews or other constrained topics. Still, for many applications they can be reasonable, especially when augmented by domain knowledge. Armed with lists of favorable and unfavorable words, simple strategies can be specified to create a favorability index. Figure 7.10 describes a straightforward strategy. The strength of this approach comes from the sheer size of the sample, rather than the actual predictive technique. Each document is classified as positive or negative based on the tf, term frequency, of positive and negative words. The overall sentiment is a ratio of positive to negative words.

If this procedure is applied to the product review example of Sect. 7.3.3.1, using dictionaries that include the high frequency words, the favorability index of 3.6 is almost equivalent to correct average user-rating of 3.4 on a 1 to 5 scale.

7.4 Hybrid Example: Independent Sources of Numerical and Text Data

In Sect. 7.3.1, we observed that *Yahoo! Finance* is a good source of historical stock price information, and the data can be readily downloaded from their website. That same website also posts the latest news articles for individual companies. Both forms of data, the price information and the news articles, could be useful in tracking stock behavior and potentially for prediction of price movements. In this example, we record the 50 and 200 day moving averages and the current and previous closing prices. To obtain this information from yahoo.com, the following command can be used, where the stock symbol is filled in as IBM, and various codes special to yahoo.com are listed, like m3 for 50 day moving average:

```
wget "http://quote.yahoo.com/d/quotes.csv?s=ibm&f=m3m4l1p"
```

With these data, a simple prediction problem could be specified from the variables described in Table 7.4, which could be computed at some fixed interval such as daily or weekly. Possibly more than one company could be considered in the same sample.

Using this strategy, sample data could be collected at fixed intervals and added to the training set. Alternatively, just the closing price data could be downloaded, and a program could compute the relevant moving averages and their relation to the current price. The result of this process is a csv file, composed of numerical data, that is ready for use by standard prediction methods.

For each stock symbol, yahoo.com provides the latest business headlines for the company from newswire services like Reuters or Forbes. These can be obtained using the following command, where IBM is the company queried:

```
wget "http://finance.yahoo.com/rss/headline?s=ibm"
```

These documents can be processed in the usual way: creating a dictionary and vectorizing the stories into a single example for a company's headline documents. The headlines and news stories might be collected simultaneously with the price data by two independent operations. The transformed data from the two sources is combined into a single example by a simple paste operation of the two vectors.

Table 7.4 Possible prediction goals

Target closes above or below previous closing price
Target closes above or below 50 day moving average
Target closes above or below 200 day moving average

Table 7.5 Hybrid data from documents and databases

sym,online,servers,gmail,r50by200,r50p,r200p,ma50,ma200,ltrade,up
ibm,4,2,0,1.088,0.981,0.902,119.091,109.445,121.35,1
goog,7,9,12,1.120,0.952,0.850,474.707,423.869,498.74,1
yhoo,10,6,5,1.043,0.916,0.878,15.8483,15.1904,17.30,1

Table 7.6 Factors for cardiovascular disease

gender

age

diabetic

smoker

systolic blood pressure

total cholesterol

good cholesterol (hdl)

waist size

family history of cardiovascular disease

(cholesterol) medication

outcome: cardiovascular disease code (ICD9 codes)

The resultant process for combining data for IBM, Google and Yahoo on a specific date is shown in Table 7.5 where word frequencies for *online*, *servers* and *gmail* are combined with various ratios and moving averages of stock prices and the predictive goal is whether the stock price went up or not.

7.5 Mixed Data in Standard Table Format

In Sect. 7.4, hybrid data were assembled from two independent sources. For prediction applications, a typical data sourcing environment is mixed data, some numerical or categorical variables, some text. The tables are well-organized. Each row is an example and the features, or columns are consistently recorded. The numerical fields can be readily separated from the text fields and each processed in the usual way. They are later readily joined in a format that fits the standard format for prediction. If the text fields are just codes, and only one can be assigned for each column, then some software packages will directly process them without any additional preparation. Alternatively, the codes can be treated as text, and then processed as words, where they are translated into binary, true-or-false features. This type of translation is essential when a column can have more than a single code per example, and the codes are not recorded in any order.

Table 7.7 is a hypothetical example of electronic medical records based on an adaption of the well-known Framingham study for evaluating cardiovascular disease risk. The factors considered are listed in Table 7.6. The treatments medications are specified as text. The column is extracted and the dictionary of Table 7.8 is used

Table 7.7 Example of electronic medical records

gender,age,diabetic,smoker,systolicbp,chol,hdl,waist,family,medication,outcome

M,50,Y,Y,142,245,31,42,Y,lovastatin 40 vitamin selenium,306.2 785.51

F,35,N,Y,137,202,47,35,N,simvastatin 5,429.2 787.0 789.0

F,26,N,N,110,179,69,32,N,atorvastatin 10 amoxicillin,428.0 668.0 675.0 685.0

M,29,Y,Y,121,221,29,38,Y,simvastatin 20,425.0

M,74,N,Y,170,280,38,45,Y,rosuvastatin 80,327.3

M,82,Y,N,155,231,52,36,N,lovastatin 40 fexofenadine,414.0 414.06

F,69,N,N,122,180,75,30,Y,lovastatin 10 estrogen,428.0

M,19,N,Y,101,195,64,31,N,rosuvastatin 5 dipivefrin,412

F,22,N,Y,119,216,53,29,Y,atorvastatin 10 besylate,414

F,47,Y,N,140,207,46,39,Y,simvastatin 10 desipramine,288.5

M,44,N,N,115,173,58,37,Y,atorvastatin 10,417 417.9

F,58,Y,N,171,254,40,40,N,rosuvastatin 20 betaxolol,413 413.0

Table 7.8 Treatment options for cholesterol	
	atorvastatin
	lovastatin
	rosuvastatin
	simvastatin

to create a vector of the four, binary treatment variables, corresponding to the four treatments for cholesterol. The cholesterol treatments are commingled with many other treatments that are ignored, as are the dosages. The first 3 digits of the (ICD9) diagnostic codes, between 410 an 450, correspond to cardiovascular disease. These codes are standard billing codes in the USA.

The result of the translation and merger of the numerical and text fields is the output of Table 7.9, which is in a standard predictive model format.

7.6 Summary

Data for automated prediction comes from many sources. In previous chapters, discussions centered on pure text mining. Here, we expand our horizons to encompass both text and structured numerical data. Initially, we review the ideal data representations for prediction using either numerical or text data. We consider numerous sources of data including databases, the web, and hybrid forms of text and numerical data. Prototypical examples of blended numerical and text data are given. Using the web as a source of data for prediction is examined. Among the examples presented of web-sourced data are downloaded scientific publications formatted in XML, stock price data and related newswire headlines. Sentiment and opinion analysis are considered with examples from online product reviews. Predictive mining of electronic medical records mining is presented as an example of mixed-data mining.

Table 7.9 Electronic medical records in standard format

gender,age,diabetic,smoker,systolicbp,chol,hdl,waist,family,a-statin,l-statin,r-statin,s-statin,risk
0,50,1,1,142,245,31,42,1,0,1,0,0,0
1,35,0,1,137,202,47,35,0,0,0,0,1,1
1,26,0,0,110,179,69,32,0,1,0,0,0,1
0,29,1,1,121,221,29,38,1,0,0,0,1,1
0,74,0,1,170,280,38,45,1,0,0,1,0,0
0,82,1,0,155,231,52,36,0,0,1,0,0,1
1,69,0,0,122,180,75,30,1,0,1,0,0,1
0,19,0,1,101,195,64,31,0,0,0,1,0,1
1,22,0,1,119,216,53,29,1,1,0,0,0,1
1,47,1,0,140,207,46,39,1,0,0,0,1,0
0,44,0,0,115,173,58,37,1,1,0,0,0,1
1,58,1,0,171,254,40,40,0,0,0,1,0,1

7.7 Historical and Bibliographical Remarks

Our presentation of text mining is from a predictive modeling perspective. Other researchers emphasize information extraction from online sources of text (Feldman and Ungar 2009). A brief discussion of sentiment analysis and some alternative approaches to analysis is found in Melville *et al.* (2009). A more detailed review of opinion mining and sentiment analysis is found in Pang and Lee (2008). Variations of sentiment analysis have been applied to massive numbers of data sources. In Leskovec *et al.* (2009), key phrases in news articles were tracked from 1.6 million media sites and blogs. An example of the commercial use of sentiment analysis is given in Rui *et al.* (2009), where comments on Twitter are used to project future product sales. An analysis of customer reviews is given in Hu and Liu (2004). The Framingham Heart Study is a continuing long-term research effort that follows thousands of patients over many years (D'Agostino *et al.* 2008).

7.8 Questions and Exercises

1. Suppose you have data in CSV format with one case per row and the class label as the first field. Use the commands in Table 7.1 to transform this into CSV format in which the first field is a unique case identifier, the last field is the case label.
2. Write a script to extract all subjects from a set of XML documents of the type in Fig. 7.3, assign them unique numeric identifiers and output a file in CSV format, one line per document, containing the subject identifiers for each document.
3. Consider the data in Table 7.3. Write a script that (a) replaces the *Close* field by the average of prior values within a 2 week window (b) adds a goal field called *Up* that is 1 if the current *Close* value is higher than the previous day's value and is 0 otherwise.

4. What is the purpose of the final *awk* command in the *getpubmed* script of Fig. 7.4?
5. Discuss how product reviews are different from movie reviews. List three aspects that make movie reviews more difficult to classify than product reviews.
6. Try the algorithm of Fig. 7.10 on some reviews downloaded from the web. How can it be adapted to properly handle negation words such as *not*?
7. In Sect. 7.4, an example was given in which several fields related to IBM's share price were extracted from Yahoo! Finance. Modify the example to also extract company information such as market capitalization and P/E ratio.
8. Discuss strategies for handling missing values in electronic medical records.

Chapter 8
Case Studies

Our approach to text mining is motivated by practical applications. However, the design and development of prediction methods often take place in a controlled scientific environment that simulates the real world. This is necessary for comparative analyses and also for unraveling the pieces of the puzzle that constitute a prediction problem. The question remains as to the appropriateness of these methods for practical use. Unlike laboratory environments, the real world is less readily controlled. Methods may need to be combined and adapted to the task at hand. User interface issues should be addressed. Practical considerations, such as resource limitations, must be acknowledged.

In this chapter, we present brief case studies applying text-mining techniques to real problems. In each study, we shall begin with a statement of the problem and then show how it can be solved. Our description emphasizes the text-mining subsystems in the solution. These case studies focus on certain generic problems that should interest readers with similar concerns. The studies may provide ideas on how text-mining techniques can be used to solve real-world problems. The solutions offered in these studies are not the only ones possible but are good illustrations of text mining in action.

8.1 Market Intelligence from the Web

8.1.1 The Problem

There are many sources of news on the Web, often taken from newswire services such as Reuters or Associated Press, that contain valuable information about companies and their products. Such information may be used, for instance, to analyze competing products and shifts in brand image perceptions. By analyzing such information from current articles, one may obtain real-time market intelligence. In this case study, the focus is on detecting critical differences between the text written about a company and the text for its competitors. The fundamental question is

S.M. Weiss et al., *Fundamentals of Predictive Text Mining*,
Texts in Computer Science 41,
DOI 10.1007/978-1-84996-226-1_8, © Springer-Verlag London Limited 2010

"What's different about the stories for this company?" To this question, one may add numerous modifiers. For example, what is different about the articles written about IBM in the current quarter versus the previous quarter? What is different about the stories for IBM versus those for Microsoft? Or, using an objective measure and a more complicated question, how did various news stories influence the relatively higher prices for IBM stock versus that of Sun Microsystems?

There are several issues that need to be addressed. What articles are to be obtained? How can they be distinguished among competitors? How can patterns be found in the articles? How should the results be displayed so that they can be understood by human analysts? Addressing these issues calls for a solution that integrates several text-mining techniques.

8.1.2 Solution Overview

The overall design of a system that solves this problem is shown in Fig. 8.1. The goal is to provide insightful market intelligence on a group of competitors. We are looking for word patterns in articles and employing text categorization techniques along with document-gathering methods. We can map the task into the following components:

1. A real-time Web crawler that monitors newswires for stories about the competitors. The fetched documents are cleaned and stored for further analysis.
2. A conditional document retriever that selects only those stories that meet the specified conditions. This can be done in the form of a query that specifies the characteristics of the document group. Typically, two groups are extracted. The documents in the extracted groups can then be labeled.
3. Text analysis techniques that convert the stories to a numerical format.
4. Rule induction methods for finding patterns in data. Although other kinds of categorizers might be useful, rules are more helpful to the analyst because the word patterns corresponding to each rule can be readily examined.
5. Presentation techniques for displaying results. These would assist the analyst in selecting interesting word patterns from the rules. The system would then highlight sections of the original stories that exhibit these patterns for specific companies.

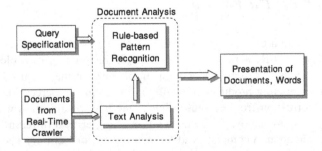

Fig. 8.1 Overview of competitive market intelligence

8.1.3 Methods and Procedures

The first task is assembling the articles for a designated group of competitors. This involves the use of a crawler that operates in real time, checking at fixed sampling intervals for the arrival of a new article for one of the competing companies. The sampling interval could be in such units as seconds or days, depending on the criticality of the requested information. To avoid repeating the same article, for each company, an index is kept of the titles and summaries of previous articles and the time of publication. It need not monitor a universal library, just the arrival of documents for one of these specified companies. Newswire agencies, such as Reuters, already index stories by company stock symbols, making it sufficient to check the library of a newswire service. These articles are posted on the Web in HTML along with extraneous content, such as advertisements. To prepare for text mining, the documents should be cleansed of irrelevant information and converted to a common format such as XML that keeps the text but eliminates the chaff. The end result is a database of documents for the competitive group in readable XML format, ready for further processing. The database may be further adjusted to contain only documents from the period of interest. For example, stories more than one month old may be deleted.

The query specification is used to label the documents. Labels are not assigned permanently but are computed based on the specifications from the analyst. We know that we will compare two or more groups of documents, but the composition of those groups can vary greatly depending on the satisfaction of various conditions for computing the labels. We may compare a company to itself over different time periods. We may compare two or more (groups of) companies over different time periods. We may compare the same company for instances when the stock price rose versus when it declined. The stories that meet the conditions are extracted from the database of documents collected by the crawler. They form a single group of documents for comparison and are assigned the same label. The process is repeated for each group specified by the analyst.

At this point, we have a text categorization problem. The data have been separated into different groups of labeled documents, all in XML format. A rule induction categorizer is invoked at this stage. The result will be analogous to reversing the usual search for documents that match a pattern using a search engine and finding instead the most likely patterns for the group of documents. First, a dictionary of k most frequent words is extracted from the data. For example, we might find the 150 most frequent words for each class and remove all the stopwords (the analyst may augment these to include words besides the usual poor predictor words such as pronouns, articles, etc.). Each document is then vectorized with binary features and the class label is attached. Then, we apply rule induction to solve the numerical classification problem.

This gives a set of rules, with each rule indicating a word pattern. Although the rules were obtained to maximize predictive performance, our objective is to review a relatively small collection of these rules to see whether they are of interest to their competitive position. So not all rules will be useful. To assist in finding interesting rules, the analyst may manipulate the list of stopwords. If the rules are found to be less than promising, some of their less satisfactory words may be added to the stopwords and the process repeated, causing the rule induction method to search for alternative patterns not containing the previous unsatisfactory words.

Finally, the analyst can retrieve the articles that satisfy the rules of interest, and the system can highlight those sections that have the indicated word patterns.

8.1.4 System Deployment

A market intelligence system such as this one is primarily for decision support and requires human interaction. The human analyst must specify and then revise the conditions of the problem, the direction to search for solutions, and the degree of interestingness of proposed solutions. Although we can expect that some of these concepts might be implicit in the preparation of data, there is sufficient flexibility, with many possibilities for the problem statement. An implementation of this system has proven to be very successful in yielding low error rates for document indexing, but it's up to the analysts to determine whether insightful or interesting patterns are found by this text-mining system.

Figure 8.2 is a snapshot of a particular implementation. It shows a straightforward comparison of news stories for IBM and Microsoft during the first two months of 2002. We see that patterns emerge relative to the retirement of the CEO of IBM and

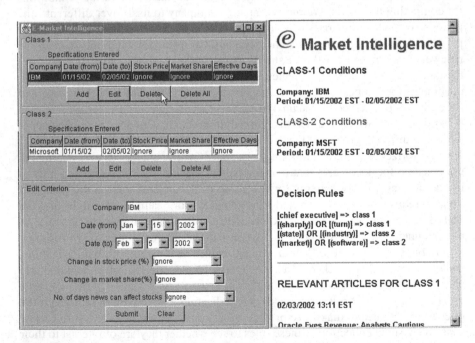

Fig. 8.2 Snapshot comparing IBM and Microsoft newswires

the states' antitrust actions against Microsoft. As an example, consider the following excerpt from a session of market intelligence analytics:

1. Starting with stories dated after September 1, 2001, crawl the newswires and collect stories for IBM, Microsoft, Dell, Compaq, and Sun. These are the designated competitors. Sample every 15 minutes and add any new materials. Clean and convert to XML. Collect share price quotes at the time of stories. Add stories to current database.
2. Specify conditions for forming groups and labels: IBM stories for December 2001 versus Sun Microsystems stories for the same period.
3. Induce rules of the form *A or B*, where *A* and *B* are conjuncts of no more than two words each. Compare the patterns in rules for the groups. For IBM, *data or sign* is a dominant pattern. Displaying the relevant articles highlights that these are mostly contracts signed by IBM for data storage and recovery.
4. Form new groups. This time, examine only IBM stories: stories from December 1, 2001 until December 10, 2001 versus stories for the rest of December 2001.
5. The resulting patterns are: (a) *services or network*; (b) *york or work*. Display articles with these patterns and highlight section containing word patterns. The first pattern appears in articles highlighting the signing of many new contracts for IBM Global Services. The second pattern is useless because *york* is just the location of New York.
6. Delete *york* and induce new rules. The resulting patterns include *sign or systems* which is added as the second pattern. Display articles and highlight words.
7. Continue analysis with new conditions: IBM market capitalization increases relative to competitors versus IBM market capitalization decreases during December 2001.

This illustrates how the analysis can continue with new queries that arise from analyzing the results of previous queries.

8.2 Lightweight Document Matching for Digital Libraries

8.2.1 The Problem

In an age of distributed and pervasive computing, mobile devices with limited capacity and restricted power are used to access server-based digital libraries. Whereas retrieval based on key words is a common way to access the contents of the digital library, we consider scenarios where users seek to retrieve documents similar to a specific document on their mobile device. For example, a technician attending to a customer problem on-site may prepare a complete description of a problem that can be stored on a laptop. That document can be indexed, and similar problems and solutions can be retrieved from a digital library at a remote server. Typically, digital libraries come with information retrieval methods that involve a text search with an input query or limited number of key words and produce an output list of potentially relevant documents. They usually require substantial storage and computing resources and are therefore server-based. This search engine style of retrieval may be adapted to use case-based reasoning methods to find documents that match a given input document, but this strategy is again less suitable for mobile devices. The problem is to devise a retrieval method that can take as input lengthy documents, give output like a search engine (of ranked matching documents), and be able to run on mobile devices.

8.2.2 Solution Overview

The solution involves the use of a lightweight document-matching method. This matcher is designed to require minimal data structures and use fast scoring algorithms to compute efficiently even on mobile computers. It exploits the structure of the input document—by focusing on certain parts of the document such as the title, key words, presence of special tags, etc.—besides obtaining the most frequent words in the document. The features extracted from the input document are matched against the documents in the library. In order for this match to take place efficiently, an offline preprocessor generates a set of data structures from the documents in the library. These data structures are: (a) local dictionaries of relevant words for each document and (b) a global dictionary of unique key words. This preprocessor is run just once at the start (if new documents are added to the library, it can incrementally update the data structures). These data structures are small enough to be cached in mobile devices. The preprocessing step is shown in Fig. 8.3.

An online matching process now uses this information to score the relevance of stored documents to an input query document. The scoring algorithm uses the count of matched words as a base score and then assigns bonuses to words that have high predictive value. It optionally assigns an extra bonus for words that matched from salient substructures in a document, such as a title, and domain-specific document

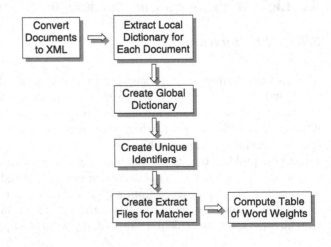

Fig. 8.3 Preprocessing for lightweight document matching

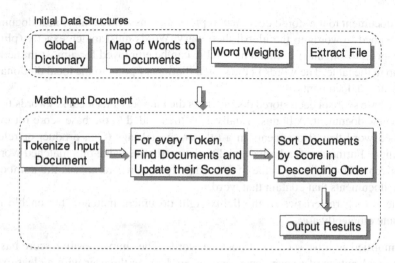

Fig. 8.4 Lightweight document matching

tags (for example, specific product or release identification associated with a document). The online process is shown in Fig. 8.4.

8.2.3 Methods and Procedures

Let's first look at the back-end process that generates the data structures necessary for the matcher to work. It uses an XML-style markup language to delineate the parts of the documents relevant to text retrieval and presentation to the end user. Two data structures are derived from the resulting file:

- A set of local dictionaries that contains the words that are relevant to specific documents. Typically, eight to ten key words are assigned for each document. The words are not unique to documents; the same word may appear in many documents.
- A pooled, global dictionary containing a list of all words that are relevant to any document. This is a unique collection of words.

The XML document contains information relevant to document retrieval that is not contained in these two data structures, such as document titles, and possibly application-specific attributes such as component identifiers. A final XML-style extract document incorporates the contents of the local dictionaries with these additional attributes. These two data structures, a global dictionary and an extract file representing a set of local dictionaries and additional attributes, are sufficient for the fast document matcher to score new documents.

The front-end matching process takes an input document and outputs a ranked list of matched documents. A special scoring function is employed to compare the input document to the stored document representations. Words in the new document are matched to words in the global dictionary. Words must match exactly (plurals are mapped to singular) so that a hash table can be employed for almost immediate lookup in the table. The words in the global dictionary point to the local dictionaries of the stored documents.

The base score of each stored document is the number of its local keywords found in the new document. A bonus, usually 1, is then added to the base score for every matched word that also appears in a title or special tags (e.g., product or release identifier). Furthermore, a bonus equal to the predictive value of a matched word is also given. The predictive value of a word is $1/num$, where num is the number of stored documents that contain that word.

The key characteristics of the lightweight document matcher that enable it to compute efficiently are:

1. Simplified additive scoring of positive words. The simple positive score has the great advantage of transparency of scoring, leaving the user with a clear explanation, in terms of identified key words, for the retrieval of a matched document.
2. Exact word match with synonyms and no stemming except for plurals. This is justified since the feature space is reduced from full indexing.
3. Reduced indexing with no document frequencies. This has the greatest impact on reducing complexity and storage requirements. The lightweight matcher uses a reduced inverted index as the basis of all matching without any additional computation.

How would this compare with a more "classical" document-matching system? Usually, document matchers compute more complicated distance measures for scoring the match. These would typically be more helpful with "recall" of documents that the lightweight matcher might be giving lesser importance. Most off-the-shelf document matchers also use full indexing, which dramatically increases the number of references to documents, the computations for matching words, and the storage requirements. In the general situation, a full index may be beneficial. But for a mobile user interested in fast retrieval, the simpler lightweight matcher is adequate.

8.2.4 System Deployment

A document-matching system such as this one is most suitable for a Java-based implementation in which the matcher is invoked from a browser. A graphical user interface for such a deployment in a help desk application is shown in Fig. 8.5. The response (the ranked list of matched documents) can be produced in HTML for easy display in a browser. The data structures used by the matcher can easily be stored on the mobile device itself and the server-based library accessed only for retrieving the specific documents needed.

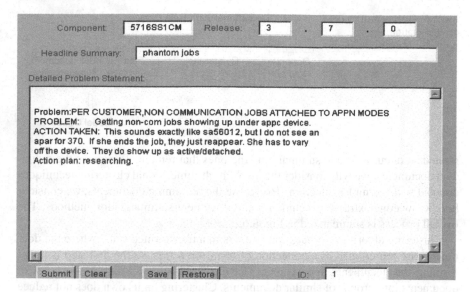

Fig. 8.5 An input interface for lightweight document matching

8.3 Generating Model Cases for Help Desk Applications

8.3.1 The Problem

In a help desk setting, customers submit problems or queries online (or via a call center) to the vendor of a product and receive solutions or answers. Each interaction can be considered a text document. Not all users of a product report unique problems to the help desk. It can be expected that most problem reports are repeat problems, with many users experiencing the same difficulty and receiving the same fix. Some of the corresponding documents may be concise, almost directly decomposing into a problem statement and solution. Others, such as those from call centers, might be almost complete transcripts of customer and service representative discussions. In such situations, individual documents may include much text that does not relate to the ultimate problem resolution. The central purpose of such documents would be simply to maintain records of a customer's interaction with a service representative.

As a result, we have a database of documents that may be ill-formed, containing redundant and poorly organized documents. Our problem is to transform the database into a concise set of summarized reports, *model cases*, which are more amenable to search and problem resolution without expert intervention.

8.3.2 Solution Overview

The heart of the solution to this problem is the use of document-clustering methods. Ideally, we would like to reduce the size of this database by eliminating the

Fig. 8.6 Overview of model case generation

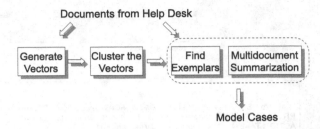

redundant documents and summarizing the ones that remain. To remove the redundant documents, we will consider the use of high-dimensional clustering techniques coupled with exemplar selection. To cleanse the remaining documents, we consider both knowledge extraction techniques and document summarization methods. The overall process is summarized in Fig. 8.6.

Clustering algorithms process documents in a transformed state, where the documents are represented as a collection of terms or words, so the first step is transforming the documents into vectors. The clustering process partitions the set of documents into groups of similar documents. Clustering on its own does not reduce the number of documents. The individual documents remain the same, only now they are grouped with similar documents. To eliminate redundancy, the final step is to select exemplars from each cluster and to produce a set of summary documents. These exemplars and summaries represent the model cases.

8.3.3 Methods and Procedures

To generate vectors, the structure of the documents is taken into account. Each document has a title. It may also have key words assigned to it. The words in the title as well as the key words are used along with the m most frequent words in the document to generate a vector for the document. A simple Boolean feature per word (to indicate its presence or absence) is usually sufficient, although other transformations may be desirable depending on the clustering method chosen. The main concept is to reduce the number of overall words that are considered, which reduces the representational and computational tasks of the clustering algorithm. Reduced indexing can be effective for these goals if performed prior to clustering. Also, the size of the documents must be taken into consideration. It is important to normalize documents. Otherwise, longer documents will tend to be similar by chance.

The clustering algorithm, much like any information retrieval system, works with this vector representation of the original documents. There are many clustering algorithms available. In principle, any of them can be used. In practice, it may be necessary to consider whether they can handle the complexity of the number of documents and number of features involved. Another criterion might be whether the number of clusters is determined dynamically or is a user-specified parameter of the clustering algorithm (in which case it would need to be tuned using validation data). How many clusters should one expect to obtain? Naturally, this depends on the nature of the documents. Because each document records a problem and a solution, and because the problems may be similar but still have critical distinctions that result in differing solutions, the number of clusters is typically relatively large, much larger than the number of clusters needed for summarization of problem types alone. How can we objectively evaluate clustering performance? Very often, the objective measure is related to the clustering technique. For example, k-means clustering can measure overall distance from the mean. Techniques that are based on nearest-neighbor distance, such as most information retrieval techniques, can measure distance from the nearest neighbor or the average distance from other cluster members.

Obtaining good clusters is very important since they designate groups of similar documents. Once clusters are obtained, exemplars can be selected from them. If the documents in a cluster are redundant (e.g., each describes the same problem and solution), then selecting one document from the cluster can be sufficient to describe the cluster. But since the clustering process is imperfect and some clusters may be large, looking at individual documents within a cluster, we may still see some variability. So it may be safer to select more than one exemplar document to represent the cluster. The same measure of evaluation used for clustering can be used to find exemplar documents for a cluster. The local dictionary of a document cluster can be used as a virtual document that is matched to the members of the cluster. The top-k matched documents can be considered a ranked list of exemplar documents for the cluster.

The set of exemplars already achieves a reduction of the original database. Alternatively, summarization might be attempted. Ideally, each cluster contains documents for the same topic. Unlike single-document summarization, summarizing documents on a common topic is sample-based and explores common patterns across many documents. The summarization can be done over all the documents in the cluster or over the exemplars of that cluster. The summarized documents can either be added to the list of exemplars or replace them. At the end of this process, model cases are obtained.

How representative are the model cases of the original database? This depends on the quality and clarity of the original documents, which, in turn, affects the quality of the clusters obtained. It is best to have the original documents stripped to problem–solution pairs before generating model cases. However, one can readily envision real-world scenarios where the documents are poorly structured. For example, consider the following possibilities:

- The customer e-mails the problem statement, and a service representative writes a solution. Because a problem–solution model is expected, and the discussants are asked to compose their thoughts in writing, the resultant document tends toward clarity and conciseness.
- The customer communicates by phone to a call center, and the service representative creates a real-time approximate transcript of their dialogue. Such documents tend to be rambling, with extraneous text.

For the first possibility, little preparation is needed. For the second possibility, additional effort is needed to extract critical sections. Knowledge-based models can be very helpful. For example, we know that the problem statement is typically at the beginning of the document and the solution is at the end. Moreover, the service representatives can be told to prefix critical sections with key words such as *action taken*. Far more helpful, and far more powerful for a self-help document, would be for the service representative to write a one-line or two-line summary of the solution. This takes a little extra time, reduces the time available for a single representative to take more calls, and is not needed when the main purpose of the document is to maintain a record of the customer's problem. However, it does have long-term gains in producing more useful model cases.

Summarization and exemplar selection are alternative approaches to document reduction, and each has its advantages and disadvantages. The exemplar approach keeps documents intact and coherent. Summarization by topic merges excerpts of many documents and is therefore more susceptible to mistakes in the extraction of the excerpts. However, the exemplar approach is more dependent on starting with cleansed documents that contain the critical sections of the document, such as the problem statement and solution. The topic summarization technique has the potential to find those sections because they will appear in many samples while discarding those sections that are specific to individual documents. So should one use exemplars or do summarization? There is no single answer that will work for all help desk applications. The best approach might well be a hybrid combination of these two approaches based on knowledge of the documents involved.

8.3.4 System Deployment

There are at least two ways of deploying such a system. Vendors may deploy it for internal use in which the system provides an overview of the problems that customers are facing. In this scenario, the system serves to assist managers and product designers interested in gathering insights from help desk queries. Such an overview may be helpful to identify trouble spots that need extra resources. For example, a computer vendor might discover, that, for their products, printer problems comprise a large percentage of customer complaints.

In a more ambitious deployment, model-case reports can increase the potential for self-help by customers. In such a scenario, a customer may file problem reports online and be automatically provided with a concise summary of the fixes that helped solve similar problems in the past. Automated procedures cannot be expected to perform these tasks perfectly. However, one can find real-world circumstances where imperfect results will still provide large benefits. Typically, for even large help desk databases, the number of model cases for the most common problems is small, perhaps no more than a few hundred. As a result, the matching process can be very efficient and provides a way to deliver a quick initial answer that might fix the customer's problem.

8.4 Assigning Topics to News Articles

8.4.1 The Problem

News services, such as Associated Press, the Wall Street Journal, Xinhua News Service, Reuters News Service, etc., provide news stories to subscribers daily. Their subscribers are newspapers, television and radio stations, Web sites, and private entities. Each of the news services provides hundreds of stories every day, although a significant number are minor variants of earlier stories. Subscriber organizations like to have these stories tagged with index terms, drawn from a fixed set for each source, so that stories can be routed to the correct desk or to a database so they can more accurately be retrieved later on. This process of assigning index tags or topic names can be accomplished by having a set of human coders read each story and assign it one or more topic codes. Besides being expensive, such a process can be inconsistent because no two coders would view the same story in exactly the same way. The problem is to devise an automated method for assigning topics (from a fixed set) to news articles.

8.4.2 Solution Overview

The solution to this problem involves text categorization. Each story is associated with a vector of features derived from the story. The features are words that occur in it, each feature having a numeric value based on the frequency of occurrence or a Boolean value indicating the presence or absence of a word. Because the potential vocabulary from a large set of news stories is a very long list, the total feature set is huge. However, a particular news story has values for only a few of the total set of features. The vector corresponding to the story is then passed through a set of binary categorizers, one for each topic code. Each categorizer gives the story a positive or negative label depending on whether the story is classified in the corresponding topic class or not. At the end of the process, all the positive labels are collected and the story is assigned the corresponding topics. The categorizers are obtained using a machine-learning algorithm that processes a set of story feature vectors for which the correct topic codes have been assigned. The overall process is summarized in Fig. 8.7. From a text-mining perspective, the two key aspects of the solution are: (a) feature generation; (b) use of multiple binary classifiers.

8.4.3 Methods and Procedures

Although the solution appears to be a straightforward application of text categorization, there are many crucial questions that pertain to its use for newswire stories. How much data should be used for training? What features are most appropriate for

Fig. 8.7 Assigning topics to newswire stories

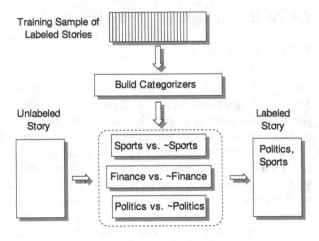

Table 8.1 Some topics in the Reuters data

Topic	Number of stories
Markets	200190
Equity markets	48700
Sports	35317
Interbank markets	28185
Annual results	23212
Elections	11532
Religion	2849
Unemployment	2136
Advertising, promotion	2084
Obituaries	844

newswire stories? Which classifier would be most suited? When should the classifiers be updated? We will answer these questions within the context of a particular study that utilizes a fairly large collection of stories made available by Reuters. Although a different data context may give different answers, the methodology used for arriving at the answers should be quite similar to the one that we outline here.

Let us first examine this Reuters sample. The collection consists of some 810,000 stories covering the Reuters newswire over a year, from August 20, 1996 to August 19, 1997. Uncompressed, it is about 2.5 gigabytes. Each story has a certain amount of associated metadata, such as creation date, publication date, source, etc. Included in the metadata are three codes: region codes, topic codes, and industry codes. We focus here on the topic codes. Some of the topics, along with the number of associated stories, are shown in Table 8.1.

For this data collection, we note that (a) there is wide variation in the frequency of topic assignment, and (b) the topic structure is partly hierarchical. For example, *Markets* is a supertopic of *Interbank Markets* and *Equity Markets*. A story classified as *Equity Markets* is necessarily also classified as *Markets*. However, not all stories are classified down to an end category. There are stories classified as *Markets* that are not in *Interbank Markets* or *Equity Markets* or any of the other subtopics. Although algorithms exist to exploit hierarchies in the classification process, we will not pursue this avenue. *Markets* and *Interbank Markets* are merely two different topics, just like *Equity Markets* and *Religion*. Because more than one topic can be assigned to a story, it is simpler to view the overall classification as a collection of binary problems.

The presence of such a large collection of training data reduces the choice of classification model to either a linear classifier or a decision tree, as both can cope well with the amount of data at hand. A decision tree classifier has the advantage that the tree can be reduced to a set of more or less understandable classification rules, which allow a classification model to be manually adjusted if desired. However, let's see what the data have to say. For our sample, linear classifiers are consistently better performers than single decision trees. A closer examination shows that this is a result of the typically highly skewed distribution of stories into newswire categories. For example, if we examine results for the three most frequent categories, three others occurring about 1/10 as often, three more categories occurring about 1/100 as often, and three occurring about 1/1000 as often, it is immediately clear that the best results are in the most frequent categories. Some less frequent categories (e.g., *sports*) do have very good categorization results and others, such as *war* and *civil war*, are fair, but many of the others are quite poor. For the less frequent categories, the decision tree gives a better result than the linear method, but even there its performance is not very impressive. Overall, based on the data, the linear classifier is superior.

A key issue for the newswire domain with time-stamped stories is how the classifier performance would decay with time. A classification model trained with documents from a time period t_1 to t_2 has a certain accuracy at the present time t_3 but presumably will have a lower accuracy at a later time t_n, decreasing with increasing n. An understanding of this decay will help determine when and how often the classifiers must be retrained. Some experimentation is necessary to assess this impact of time for a specific sample. We shall illustrate the nature of the experimentation involved. Since we have data for one year, we can partition them into 12 groups, each corresponding to all stories appearing during a month. We use the twelfth month as our test data and train classification models based on each of the 11 preceding months, the previous 2 months, the previous 3 months, the previous 6 months, and the preceding 11 months (i.e., all the available training data). The objective is to see how performance varies with the different windows into the past. In doing this kind of partition, care must be taken to ensure that there are enough data to train the classifiers. For the available data, the number of documents per month is 67,500, with some seasonal variation. For example, the number of stories for the first month is 62,935, for the last training month 66,193, and for the test set 69,626. For these data, the breakdown into months is justified. Looking at the performance of linear models, we observed that the model built on month 1 data had a performance just slightly lower than the model built on the whole collection. The implications of this in an operational environment are significant and comforting. At least for this kind of data, frequent model updating does not appear to be necessary if the original linear model was built on sufficient data.

Another implication of the experiment above is the amount of data needed for training. For the linear model, it would seem that using just one month of data is just as good as using the full 12 months of data. In contrast, the situation for decision tree models is not as encouraging—using more historical data provides a clear improvement over using only one of the preceding months. Clearly, for these data, linear classifiers are preferable unless there is some overwhelming advantage to be gained by having a transparent rule-based solution instead.

For newswire stories, the text is usually grammatical and case information is best retained in the tokens and features extracted. The feature sets can range from all possible tokens, to tokens not including numbers, to tokens that had been stemmed by an inflectional stemmer. Which is most suitable? Auxiliary experiments can be conducted to determine the effectiveness of different feature sets. For the Reuters data, feature set sizes ranged from almost 400,000 to 1000. We obtained quite reasonable results with 5000 features. With more than 5000 features, the additional performance gains were much smaller. Stemming was somewhat helpful for decision trees and when story titles were not considered as special features. But, in general, stemming was not worth the extra computational effort. Using titles as features was clearly beneficial. This is not unexpected. Newswire articles tend to rely on high-information titles to give the reader a quick feel for the topic.

An examination of the categories with poor performance sheds some light on how performance is affected by the nature of newswire stories. For example, it appears that recency matters less for less frequent categories. A possible explanation is that news topics tend to occur in bursts, so training data for a particular category is localized and may not have appeared recently. Although plausible, it would take a detailed study to verify that this is indeed the case.

Another issue is that although some topics seem to be inherently cohesive and it would seem that it should be easy to build high-performance classifiers for them, this does not always happen. As an example from the Reuters data, we looked at the rule set generated by the decision tree classifier for the *religion* topic, did an analysis of classification failures for the first 40 or so errors, and have the following comments:

- As is often the case in newswires, stories that are substantially identical are sent over the wire numerous times. Some of these stories are summaries with several topics. If one of these is related to religion but is only a brief mention, the classifier will likely not find it, and because the story appears several times, what is really a single mistake appears as multiple mistakes.
- Some stories may have religion as only a subtopic of some other theme. For example, there are stories in the collection regarding human rights in China in which religion is one of the rights. This story and minor variants are also repeated, leading to multiple errors when not classed as *religion* by the classifier.
- In some cases, the classifier did not generate a rule that would have covered a number of cases. For example, there is no rule regarding multiple occurrences in a story of the word *religion*, which humans might expect to be reasonable.

Apparently, many stories that are not categorized as religion also contain words typically associated with religion. In fact, only about half the occurrences in the test set of the word *religion* are in stories categorized as *religion*. The remainder are in a number of different categories, the most frequent being *international relations* and *domestic politics*.

This illustrates the difficulty in predicting the performance of topic classifiers based on human intuition. It is best to let the data speak for themselves.

In all these experiments, the three usual metrics are useful for measuring performance: precision, recall, and F-measure. The F-measure is useful for simple comparisons among alternatives. When the F-measures are comparable, the precision and recall can be examined. When examining the performance of a system that consists of a large number of classifiers, it is common to compute the performance of each classifier and then compute their *microaverage*. For example, we compute the number of true positives, false positives, and false negatives for each of the 103 classifiers needed for the Reuters data. We sum them to get the total number of true positives, false positives, and false negatives. Microaveraged precision, recall, and F-measure for the overall system are computed over these totals.

We can draw several conclusions about classifying newswire stories from our experience with the Reuters data:

- The key characteristics of newswire data are: (a) big data; (b) many overlapping, possibly hierarchical topics; and (c) time-stamped stories.
- Even with the much larger feature space and sample sizes, the simplest methods can be quite effective. One may not need to do any special processing to create a dictionary. Words with numerical content do not necessarily have additional predictive value and can possibly be ignored.
- It is clear that the amount and recency of training data needed to build a classification model for newswire texts are dependent on the model. A linear model is much less sensitive to the recency effect and requires less training data.
- Single decision tree or decision rule classifiers can be more accurate for less frequent (rare) topics, especially if they take into account the highly skewed populations of such categories.

8.4.4 System Deployment

The obvious deployment of such a system is as a filter for an incoming stream of newswire articles. As stories arrive at the subscriber, they are automatically tagged by the system and, depending on the topics assigned, distributed to the appropriate people or indexed into a database. The advantage of such a deployment is timely access by the subscriber and less resources for monitoring the incoming stream. It may also be usefully employed as a filter for outgoing newswire articles, in which case the distributor maintains the list of topics and the system helps deliver articles that are labeled with more consistency than manually labeled articles.

Depending on the confidence in the system, its topic assignments may be manually reviewed before processing. Since the story has already been assigned topics, the manual review can be much faster than labeling the story from scratch. Alternatively, its performance may be reviewed periodically with mechanisms in place to collect feedback from the recipients.

Procedures need to be in place for maintenance of the system as well. The recipients can be asked to review the labeling of the stories they receive, and their feedback can be stored as the "correct" labels for the stories. In this manner, a collection of labeled stories can be grown. Periodically, or when performance deteriorates, such collections can be used to retrain the classifiers in the system. Retraining may also be necessary when new topics are introduced.

8.5 E-mail Filtering

8.5.1 The Problem

With the widespread use of e-mail for communication, the volume of e-mail that a person receives in a day can be quite formidable. Without any tools to manage and process the incoming messages, one could easily spend all of one's time just processing e-mail, and it's easy to miss more urgent messages buried in a mass of junk mail. Especially useful would be a tool that organizes incoming messages into a set of folders. The tool needs to be highly configurable—each user (or even the same user at different times) will have different sorts of folders and different kinds of messages. Of special interest is the class of *spam* messages, unsolicited messages sent automatically by mass-marketing agencies. It would be useful if the tool could also distinguish spam from nonspam messages. The tool will have full access to the content of the messages as well as other characteristics (such as message date and time, sender details, etc.). Designing this tool is the well-known problem of *e-mail filtering*.

8.5.2 Solution Overview

As can be expected for an issue of such importance, many approaches to solving this problem have been proposed. Every modern e-mail client includes some form of e-mail filtering. We shall examine one such solution available in a version of the open-source Mozilla e-mail client. Most other solutions have a similar structure. Mail arrives in a special folder called the *Inbox*. The e-mail filter is typically applied to messages in this folder. There are two distinct parts of this filter:

- A set of rules specified by the user. These rules are applied by the system to messages in a folder such as Inbox and can be moved (or its attributes modified) as directed by satisfied rules. The user specifies the rules, which rely on

Fig. 8.8 Overview of e-mail filtering in Mozilla

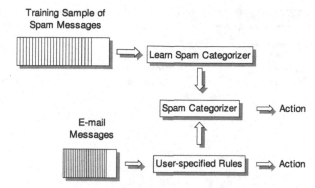

string-matching expressions (e.g., regular expressions) to check the contents of a message. The rules also exploit the structure of the message. For example, one can distinguish among message fields such as sender or subject.

- A trainable classifier that identifies all the spam messages. All messages identified as spam are put into a designated junk folder (which the user can review at leisure to ensure that nothing important has been deleted). To train the classifier, facilities are available for labeling messages as spam or nonspam.

The overall process of e-mail filtering in Mozilla is shown in Fig. 8.8.

8.5.3 Methods and Procedures

The user-specified rules in Mozilla have an expressive structure that allows the user to create powerful but intuitive rules to classify messages in a folder such as Inbox. If a rule is satisfied, the action to be taken can be as simple as moving the message to another folder, or it can involve changing attributes of the message (e.g., its priority). In each rule, the user can specify atomic conditions and then specify whether a conjunct or disjunct of these conditions should be satisfied. The atomic conditions themselves can be focused on a specific field of the message (e.g., the sender field, the body of the message, the subject, etc.). In this version of Mozilla, simple strings can be specified for matching purposes. User-specified rules represent an easy and intuitive way for users to manage their messages. For example, it's easy to add a rule that if the sender of a message is one's boss, then its priority should be increased. More commonly, messages that match a certain pattern can be simply sent to the trash folder. An example of this is shown in Fig. 8.9.

The spam classifier in this version of Mozilla is a Bayesian model. This essentially consists of a large table of estimates of the spam probabilities of words in the message. There is a clear distinction among the impacts of different errors. A spam message misclassified as nonspam (false negative for a spam identifier) is a less serious error than a nonspam message identified as spam (false positive). Ideally, the spam classifier should generate zero false positive errors even at the risk of generating more false negative errors.

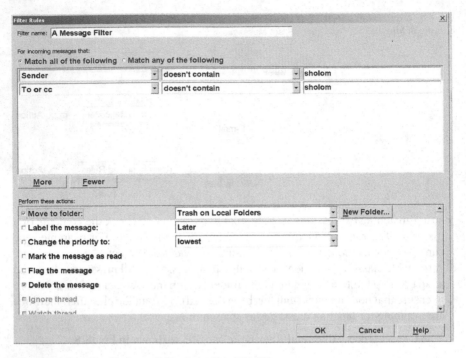

Fig. 8.9 A user-specified e-mail filtering rule in Mozilla

To use the spam classifier, a message is tokenized and the 15 most interesting to-
kens/words are found. Interestingness is measured within Mozilla by how far their
spam probability is from a neutral .5. Their associated probabilities are taken from
the Bayesian model, and an estimate is computed of the probability that the message is spam. If this probability is greater than a certain threshold (in this version
of Mozilla it is 90%), the message is classified as spam; otherwise it is nonspam.
Because of the cost of tokenizing, table lookup, and probability computation, the
spam classifier is used after the user-defined rules have been applied. At this stage,
hopefully there are fewer messages to process.

The spam classifier is built using a corpus of messages in which each message is
labeled as spam or nonspam by the user. Each message is viewed as a text document,
including headers and embedded HTML, Javascript, etc. All parts of the message
that can be viewed as text are considered. The text is tokenized and tokens that are
all digits or HTML comments are ignored. Case differences are also ignored. For
each of the two categories, spam and nonspam, two large tables are computed with
the words in messages of that category, their corresponding counts, and the number
of messages. From these two tables, it is straightforward to compute a third table
of probability estimates for each word. This third table is the Bayesian model. To
compute the probability estimates, the particular formula used within Mozilla is an
empirical one and is only one of several possibilities. The crucial aspect is that it
biases the probabilities to avoid false positives as much as possible.

8.5.4 System Deployment

E-mail filters are typically deployed in e-mail clients. They can result in significant productivity boosts in the workplace by assisting employees in managing and processing their e-mail. Such filters can also be used to route incoming messages to the appropriate person. For example, a customer service center, may have only one e-mail address but several representatives.

Specifying the rules is typically done through a user interface that ensures that the rules are syntactically correct. An interface that caters to several levels of user expertise would be useful. Casual users may, for example, prefer to duplicate and modify predefined rules. Intermediate users may want to define their rules from scratch. Advanced users may want to tweak collections of rules by reordering them or by examining historical logs of filtering behavior.

The advantages of spam identification are obvious at both the system level and the user level. The main issue with spam identification is the constantly changing nature of spam, which requires frequent retraining. Crucial to effective deployment is the user interface for specifying the training cases. If it's too cumbersome, users may not be inclined to use it or may end up with weak classifiers trained on insufficient or outdated data. Although the Mozilla client includes a Bayesian classifier, in principle a rule-based classifier would be more compatible with the user-specified rules. It would allow the user to manually tweak the generated rules and would provide the user with transparency of the spam identification process.

8.6 Search Engines

8.6.1 The Problem

Search engines were discussed earlier in the context of information retrieval. The problem that search engines address is a special kind of information retrieval. The user specifies a *search query* that consists of relevant words. The problem for the search engine is to output a *ranked* list of Web pages that match the user specification. The most relevant pages must have a higher rank. Part of the problem is to design a query language that allows users to specify precisely the pages they wish to retrieve. Obviously, the language has to be easy to use. The queries themselves are expected to be "general purpose" (i.e., not specialized for any specific subject or topic). They might even be complete sentences. However, queries are expected to be composed of few words.

8.6.2 Solution Overview

There are many search engines, and each does things a bit differently. Here we shall describe a solution inspired by the search engine Ask.com. The search engine

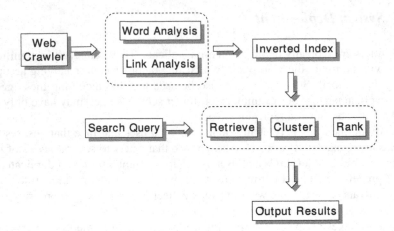

Fig. 8.10 Overview of search engine structure

structure is summarized in Fig. 8.10. There are typically three aspects to handling general queries from users:

- First, an index must be created that links words to documents. This index creation process usually involves a Web crawler that retrieves documents from the Web. These retrieved documents are subjected to link and text analyses to create the inverted index that links words to documents. This index is central to the search algorithms. Since the set of documents on the Web is constantly changing, the index must be maintained and updated as necessary.
- The front end of the search engine accepts queries from the user. The query is parsed and a set of relevant documents is retrieved. These documents are clustered into groups that can further be labeled based on the common words within each group and the words in the search query. Finally, the documents can be ranked based on the links.
- The ranked list can now be presented to the user as the output of the search query. Besides this ranked list, other useful information may also be provided. For example, the labeled clusters may be offered as a way for users to refine their search. Also, the most authoritative documents within the clusters may be provided to the user as additional resources.

8.6.3 Methods and Procedures

The query language used by search engines is quite simple. Key words are provided as a list. A meta-language is available that allows the introduction of logical operators (AND, OR, etc.). Users may specify whether case information should be considered or not. The query processor also includes a spell-checker that allows it to auto-correct typographical errors made by users. Additional advanced search tools allow the user to focus the search query by diverse criteria such as geographical region, domain, date, language, location of search terms in page, etc. These serve as filters with meta-information that can drastically reduce the search space.

The index is crucial for search quality (the search engine cannot find what it has not indexed). However, as the index size grows, the search process takes more time and more complex techniques are needed. The size of the index can be controlled somewhat by limiting the size of the input text. When considering documents for indexing, it may be truncated to the first 100 words, for example. The justification is that the most relevant parts of a document are typically at the beginning of a document. Auxiliary information about the word's position and frequency in the document, and also information about links, may be computed and stored during the assembly of an index.

The inverted index allows easy retrieval of documents. Ranking these documents can be done simply using link analysis (as described in Chap. 4). Alternatively, the relevance of a document to the search query can be measured even more precisely by giving higher priority to links from *authoritative* documents. Instead of determining a document's authority by its popularity among all documents, it can be determined more precisely by examining its popularity among documents on the same subject. This requires first forming clusters of the documents. These clusters can also be labeled using common identifying words in a cluster and the query key words. Once the ranks are determined, the documents can be sorted by relevance.

A by-product of this process is that it also clusters documents into subjects that are hierarchically organized, thereby allowing search results to be seen in context. This context allows the user to further refine the search. Since the authoritative documents can also be identified, they can also be provided to the user as additional resources.

8.6.4 System Deployment

Search engines are designed for retrieving pages from the World Wide Web. This is their most widely observed form of deployment. However, by restricting the space of documents they search, search engines can also be deployed usefully in other settings. For example, corporate Web sites with a large number of pages can benefit from a search engine that allows visitors to quickly retrieve relevant pages. Search engines can be deployed by clients of self-help applications to search a knowledge base of static documents.

Increasingly, search engine technology is being accessed from *toolbars* that integrate access to the search engine databases with the browser itself. This allows the user to perform a search from any Web page. Toolbars also allow the user to customize the presentation of results. For example, users may want to highlight the search key words in the retrieved documents, or they may want links to the current document's text.

Search engines may also be deployed such that they separate the retrieved links into several categories. For example, some search engines sell advertisers placement in a *Sponsored links* section in which highly relevant sponsored results are displayed.

More recently, the emphasis has shifted to improve the user experience by organizing results into classes that help the user efficiently refine the search query. List of search suggestions are often provided in real-time as the query is entered. Almost all modern search engines provide a *related links* section to help the user further refine the search. The *related links* section allows the search engine to also be used for document matching. Clicking on this button would make the current page a search-query string and return highly relevant documents.

Results may also be organized into search categories that directly reflect the motivations for the search. For example, the search engine Bing markets itself as a core provider of the categories Shopping, Health, Local and Travel. As an example, Fig. 8.11 illustrates the presentation of search results by the Bing search engine.

The objective of searching is to answer a question and many search engines attempt to do just that by giving *instant answers*. Usually this involves extracting the

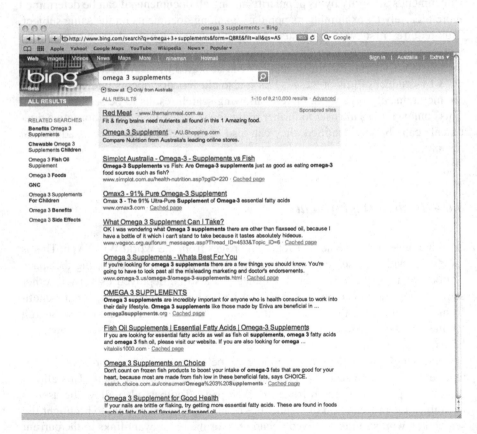

Fig. 8.11 Screenshot of the Bing search engine

desired information from relevant pages. For example, a query about flight status can be answered directly. Answering questions can involve some computation as well. For example, the result of a mathematical formula. The engine Wolfram-alpha is an instance of this emerging trend.

8.7 Extracting Named Entities from Documents

8.7.1 The Problem

Named entity recognition was discussed earlier as a key task in extracting information from unstructured electronic text. This topic was a central theme in the message understanding conferences (MUCs). More recently, it has also received attention in the form of *shared tasks* in the CoNLL conferences. These shared tasks have the following format: (a) annotated training and validation data are provided to all participants (hence the name *shared task*); (b) the participants design and tune their systems; and (c) these systems are then tested on unseen data and results are compared and analyzed. In this case study, we examine the shared task in CoNLL-2003. The data consist of annotated tokens for English and German text. The text is from newswire stories. Every token is classified as one of these nine categories: O, B-ORG, B-PER, B-LOC, B-MISC, I-ORG, I-PER, I-LOC, and I-MISC. Besides annotated data, some dictionaries are also supplied. The goal is to build a robust named entity recognition system for such data. The methodology should be applicable to a number of languages.

8.7.2 Solution Overview

There are many solutions proposed for this problem. The core of the one we will describe is a robust linear classification system. It involves the use of a very large number of features. The features can be grouped into levels of linguistic sophistication. The simplest ones are token-based and are available for many languages. More complex ones are increasingly language-specific. Feature values are extracted from the training data and nine linear classifiers are constructed—binary classifiers, one for each of the nine classes. Validation data are used to assess the impact of the features, and the ones that optimize system performance are retained.

For new text, features are extracted for each token, which is then scored by the nine classifiers. The token is classified based on these scores. The class with the highest score wins. After all the tokens for a sentence are classified in this manner, some smoothing is performed to remove any inconsistent classifications of individual tokens. The overall process is shown in Fig. 8.12.

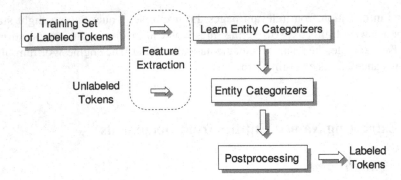

Fig. 8.12 Labeling tokens as named entities

8.7.3 Methods and Procedures

General techniques for named entity recognition have been discussed in Chap. 6. We follow the general approach outlined there and treat the problem as a sequential token-based tagging problem. The goal is to predict the class label associated with each token based on features extracted from the text. Since the features are crucial for prediction, we will discuss them in more detail.

The basic linguistic features are all aligned with the tokens. Specifically, we consider features listed in Table 8.2. These features are represented using a binary encoding scheme where each component of the feature vector corresponds to an occurrence of a feature.

After studying many different feature combinations on the English development set, the ones shown in Table 8.3 were deemed interesting. The combinations have been listed such that there is a distinct incremental improvement with an increasing numerical Experiment ID. To give a rough indication of the scale of improvement, the F-measure on an independent test set ranged from 83 to 92. More details can be found in the references cited in Sect. 8.10. We now analyze the impact of these combinations of features.

There was only a small performance difference between Experiment 1 and Experiment 2. This implies that tokens by themselves, whether represented as mixed-case text or not, do not significantly affect the system performance.

Experiment 3 showed that even without case information, the performance of a statistical named entity recognition system can be greatly enhanced with token prefix and suffix information. Intuitively, such information allows us to build a character-based token model that can predict whether an (unseen) English word looks like an entity type or not. The performance of this experiment was comparable to that of the mixed-case English text plus capitalization feature in Experiment 4.

Experiment 4 suggested that capitalization is a very useful feature for mixed-case text and can greatly enhance the performance of a named entity recognition system. With token prefix and suffix information that incorporates a character-based entity model, the system performance is further enhanced in Experiment 5.

Table 8.2 Features for named entity recognition

Feature ID	Feature description
A	Tokens turned into all upper case in a window of ±2
B	Tokens themselves in a window of ±2
C	The previous two predicted tags and the conjunction of the previous tag and the current token
D	Initial cap of tokens in a window of ±2
E	More elaborate word type information: initial cap, all cap, all digits, etc.
F	Token prefix and token suffix (up to length four)
G	POS annotation provided in the shared task
H	Syntactic chunking annotation provided in the shared task
I	A number of dictionaries from different sources: location, person, and organizations

Table 8.3 Interesting feature combinations for named entity recognition

Experiment ID	Features used
1	A+C
2	B+C
3	A+F
4	B+C+D
5	B+C+D+E+F
6	B+C+D+E+F+G+H
7	B+C+D+E+F+G+H+I

Up to Experiment 5, only very simple token-based linguistic features have been used. Despite their simplicity, these features give very significant performance enhancement. In addition, such features are readily available for many languages, implying that they can be used in a language-independent statistical named entity recognition system.

In Experiment 6, we added part-of-speech and text-chunking information. They only lead to a relatively small improvement. This is because most part-of-speech information has already been captured in the capitalization and prefix/suffix features. The chunking information might be more useful, though its value is still quite limited.

By adding additional dictionaries, we observe a small but statistically significant improvement (Experiment 7). Some of these dictionaries are provided by the shared tasks. One may also construct dictionaries from various sources, such as certain Web sites that contain lists of cities, countries, US states, etc.

Clearly, the construction of extra linguistic features is open-ended. It is possible to improve system performance with additional and higher-quality dictionaries. Although dictionaries are language-dependent, they are often fairly readily available, and providing them does not pose a major impediment to customizing a language-independent system. However, for more difficult cases, it may be necessary to provide high precision, manually developed rules to capture particular linguistic patterns. Language-dependent features of this kind are harder to develop than dictionaries and correspondingly pose a greater obstacle to customizing a language-independent system. We have found that such features can appreciably improve the performance of our system. A related idea is to combine the outputs of different systems. Fortunately, our experiments indicate that special-purpose patterns may not be necessary for quite reasonable accuracy.

The performance of a named entity recognition system is sensitive to the data source and language. For example, German can be a more difficult language for named entity recognition. However, features listed in this case study are useful for these different languages as well.

8.7.4 System Deployment

A named entity recognition system can be deployed in at least two ways. The named entities can be extracted from the input text and provided as output. The user may decide how to use the named entities. For example, some of them may be the basis for indexing documents in a database. Another way of deploying the system is within a document browser. The named entities recognized by the system are annotated for rapid perusal. For example, they might be highlighted, as in Fig. 8.13, or perhaps color-coded.

Typically, for a different language, a different classifier would be necessary. The methodology for obtaining the classifier would be the same: obtain training and validation data; extract features; train the classifier; and tune the classifier using validation data. The features extracted may differ somewhat across languages. For the CoNLL-2003 data, we found that system performance was enhanced significantly with simple local linguistic features and more sophisticated features, although helpful, yielded much less improvement than might be expected. Since the simple features are available for many languages, it suggests the possibility of setting up a language-independent named entity recognition system quickly so that its performance is close to a system that uses much more sophisticated, language-dependent features.

8.8 Customized Newspapers

8.8.1 The Problem

The Web has an abundance of news sources. These sources often update in real time with breaking news. Similar events are sometimes reported from different perspectives. An individual tapping into these sources takes on an editorial role in deciding

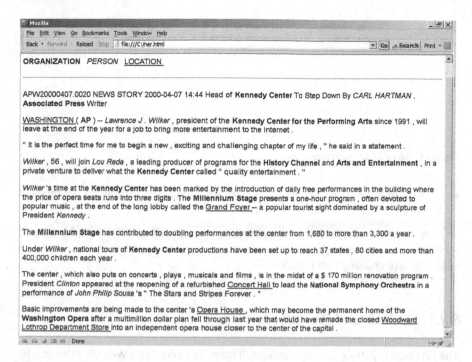

Fig. 8.13 Screenshot of a document with highlighted named entities

what is relevant or useful. With experience, a person may get better at figuring out which sites to access for which stories. Still, considerable time is spent simply visiting different sites and sifting through repetitive material. The problem is to devise a system that fully automates this editorial process, composing a customized newspaper from multiple news sources.

8.8.2 Solution Overview

An overall solution is shown in Fig. 8.14, which summarizes the different components involved. The key components of the solution are:

- A Web crawler that visits news Web sites and retrieves documents of interest.
- A categorizer that classifies articles into several key topics known in advance. For instance, the categories may be entertainment, government, finance, sports, and international.
- A document clusterer that clusters the documents within each category based on their similarity with each other. It may also take into account other attributes of the documents such as time, source, labels, etc.
- An exemplar selector that selects key articles within each category and selects other relevant articles on the same subtopics.

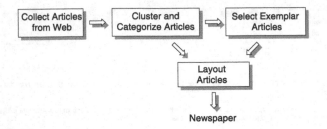

Fig. 8.14 Automating customized news

- A layout program that uses the relative importance of the various documents and categories to decide on the overall layout of the newspaper.

Note that several variations of this solution are possible. For example, a more sophisticated version might involve merging the articles together. Another version might involve creating new headlines.

8.8.3 Methods and Procedures

The first step is getting the news articles. There are many news Web sites that can be visited, and articles can be retrieved using a Web crawler. One issue in this step is cleaning up the articles and removing extra HTML links, Javascript, advertisement links, etc.

Articles are then vectorized. The vectors are the basis for many of the subsequent steps. For vectorization, it is usually sufficient to use around a few hundred terms (with the highest tf-idf weights, for example) to describe the articles. Stopwords may be eliminated, and synonyms and spelling errors should be taken into account. When documents vary greatly in length, normalization provides a common basis for comparison and prevents inappropriately high similarity measures for longer documents.

Once document vectors are obtained, they can be categorized and clustered. Besides the usual terms that capture document content, additional information may be used such as the time stamp of the document. Instead of comparing document vectors directly, it's preferable to work with topic vectors instead. Topic vectors can be created in advance for the usual news categories (such as World, Sci/Tech, US, Sports, Business, etc.), and comparisons with such topic vectors can assist in categorizing and clustering. For instance, one can create a *topic affinity vector*, that measures the similarity of a document vector to the various topic vectors. Then, instead of comparing two document vectors directly, one can compare their topic affinity vectors instead. The more similar these vectors, the more likely that they are of the same topic.

How are topic vectors created in the first place? They can be created from Web directories, which consist of articles categorized into various topics. Using the terms from all the documents for a topic, a topic vector can be generated.

This requires that topics be known beforehand. Typically, there are several key topics of interest, such as entertainment, government, finance, and sports. Names of new topics can also be generated by examining the words in a cluster, ranking them by weight, and taking the top few words as descriptive of the articles in that cluster.

Documents for each category can be filtered for relevance and user preferences. Those deemed unsuitable can be discarded. When many documents are similar, a few may be selected as exemplars. In this stage, the document sources may also be taken into consideration. An article from one news source may be given a higher weight than that from another source.

Once the articles are selected, they need to be presented in Web pages. Automated layout can be quite tricky if done from scratch. But default templates may be used as a starting point and the layout modified based on the importance and newsworthiness of the articles. Some articles may have pictures attached to them, and these too need to be considered by the layout algorithm. Unlike the layout on fixed-size pages, layout for Web pages is more flexible, with the primary concern being the hierarchical organization of links to the articles.

8.8.4 System Deployment

Automated editing allows for huge savings in manpower and increased responsiveness to evolving news. If a wide range of news sources is considered, the system allows the bringing together of sharply different perspectives on the same event.

Customized newspapers can be deployed in very diverse scenarios by including user preferences. In the simplest form, users can specify news categories of interest. For example, a user may be interested only in sports and entertainment news. Or one may customize a regional news section based on the region of interest. More complex personalizations may affect the delivery of the news itself. For example, one may specify e-mail delivery of hourly news updates on certain topics. At an extreme, customized news may take the form of a market intelligence report in which articles about specific companies at the time of certain events (such as sharp stock price changes) are collected together.

8.9 Summary

In this chapter case studies for text-mining applications are presented. Each case study is examined for the following characteristics: (a) problem description, (b) solution overview, (c) methods and procedures, and (d) system deployment. The following applications are reviewed: market intelligence from the web, lightweight document matching for digital libraries, generating model cases for help desk applications, assigning topics to news articles, e-mail filtering, search engines, extracting named entities from documents, and customized newspapers.

8.10 Historical and Bibliographical Remarks

As with most data-mining case studies, key details of deployed systems are closely held trade secrets. The systems for which more details are known have roots in published work from universities or research labs, but the crucial reworkings for successful deployment often are not made available. Not many companies are interested in helping their competitors match their system's performance!

A market intelligence system is described in Weiss and Verma (2002). An evaluation of the lightweight document-matching algorithm is provided in Weiss *et al.* (2000b). Generating model cases by document clustering is discussed in Weiss *et al.* (2000a). Details of text categorization experiments with the Reuters data can be found in Damerau *et al.* (2004). The version of e-mail filtering described is from the open-source Mozilla v1.3; the latest version can usually be obtained from http://www.mozilla.org. Its spam classifier is based on an algorithm by Graham (2003) (see also http://www.paulgraham.com). Most well-known search engines are based on proprietary techniques. An overview of the CoNLL-2003 shared task, with details of performance results, is available in Sang and De Meulder (2003). A detailed analysis of the impact of features can be found in Zhang and Johnson (2003).

Customized newspapers based on user preferences have been around for quite a while. Almost all online newspapers allow some degree of customization. The simplest instance of this is where the weather report can be for a user-specified city. Mostly though, the customizations are such that they restrict or selectively show parts of the full newspaper. The content and layout are still mostly done by humans. A system that is organized somewhat along the lines of our case study is Newsblaster, developed at Columbia University (McKeown *et al.* 2002). A view of Newsblaster in action can be found at http://newsblaster.cs.columbia.edu at the time of writing. Google News (http://news.google.com) is an incarnation of a fully automated newspaper in which the editing and news selection are done by programs. It uses proprietary procedures to compose a meta-newspaper for a fixed number of topics with exemplar links chosen as key articles and secondary links to other articles on the same topic.

8.11 Questions and Exercises

1. What kind of rule-based system would be most appropriate for the market intelligence application?
2. Why bother with lightweight document matching when search can be performed from smartphones?
3. For classifying news articles, how might the class hierarchy be exploited in the classification process?
4. How might one create an automated newspaper with a bias? For example, suppose you want your newspaper to cover events from a conservative (right-wing) viewpoint.

Chapter 9
Emerging Directions

Our principal objective is predictive text mining. The wealth of research literature for text mining encompasses a much wider range of topics than is presented here. At the same time, the research literature deals with each of these topics in great depth, describing many alternatives to the approaches that we have selected. Our description is not a comprehensive review of the field. We have used our judgment in selecting the basic areas of interest and the fundamental concepts that can lead to practical results. For example, thousands of papers have been written on classification methods. We picked our favorites for text mining.

Perusing the research literature, one sees a number of topics that we have not covered. Many of these may be quite interesting and of practical importance, but they are not special to text mining. They are general issues of data mining. Other topics are clearly related to text mining but are only weakly related to prediction.

We don't want to ignore these other topics. Yet, we will not cover them prominently because of the aforementioned rationale. As before, we make our selections of important topics. Although text mining is far from a mature field, some new areas of application are emerging. Several years from now, techniques for applying these concepts will be routine. For others, progress may be made, but they will remain on a research agenda.

9.1 Summarization

Now that documents are in digital form, they are amenable to transformations that are not expected for paper documents. You are presented with a lengthy document or a collection of documents on the same topic. Instead of reading these documents, you might request a summary. You may eventually read the full version. First, you want to see the short version. How long is the short version? Surprisingly, there are techniques that produce customized summarizations. You specify the size, for example 10% of the originals, and the programs will comply.

A principal technical approach to summarization is closely aligned with clustering. A cluster consists of similar documents. The clusters are considered classes

S.M. Weiss et al., *Fundamentals of Predictive Text Mining*,
Texts in Computer Science 41,
DOI 10.1007/978-1-84996-226-1_9, © Springer-Verlag London Limited 2010

Fig. 9.1 Summarization by
exemplar selection

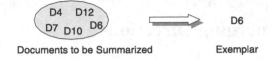

Documents to be Summarized   Exemplar

Fig. 9.2 Summarization by
merging documents

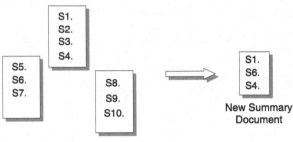

Documents to be Summarized

of documents having the same topic. We could envision several situations where a
summary might be useful:

- single document
- multiple documents with the same topic
- an automatically assembled cluster of documents.

The common theme is that one or more documents are presented, and from these
documents a summary is produced. Don't let your imagination run wild thinking that
a program must rewrite the originals. Our summary will be word-for-word sentences
extracted from the given documents. The task is one of selecting the right sentences.
This can be done by invoking a process that we described in Chaps. 5 and 7 for
summarizing the results in a cluster. Figure 9.1 describes that situation, where a
cluster is summarized by one (or more) of its constituent documents. The extract is
selected to be representative of the cluster, a summary of the shared topic.

Figure 9.2 illustrates the extended task of producing a summary by selecting sentences from the given documents. Instead of selecting a single representative document, the summarizing program extracts sentences from the documents, merging
them into a single summary.

How did we previously select the exemplar for a clustering procedure? Let's
assume we used k-means. We then have a mathematical summary of the cluster, expressed in terms of the means, such as the tf-idf of each word in the dictionary. One
simple procedure is to select the document that is most similar to the virtual document represented by the mean vector. We can use a similar procedure for selecting
sentence extracts for summarization. Figure 9.3 gives the steps for one prominent
summarization method. Using similarity measures relative to the mean vector, sentences are added to the summary. These sentences are not modified and maintain
their relative positions within the original document. Multiple documents also keep
their order relative to their time stamp, with extracts from the earliest time stamp
appearing first. Redundancy is checked by pairwise similarity scoring. Instead of

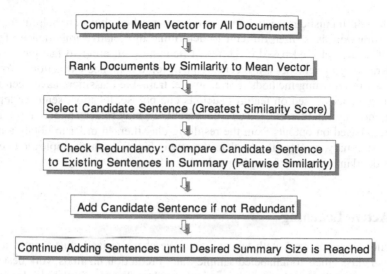

Fig. 9.3 A summarization algorithm

a pure similarity score, a weighted score that includes other factors may be used. These include the position of the sentence in the document and the length of the sentence. For example, sentences appearing at the beginning of the document may be weighted higher than those in the middle. Given a percentage threshold, sentences are added until the threshold is exceeded.

This procedure can be applied to multiple documents on a single topic or just one long document. The resulting summary is a set of sentences extracted from the full documents. Originally, summarization methods were developed using a fairly detailed linguistic analysis. But once again, we see that automated methods using procedures evolving from clustering and classification can be effective in producing a satisfactory result while circumventing any need for a deep linguistic understanding of text.

A related summarization method extracts sentences using clustering. The idea can be described as follows. We partition sentences from a document (or multiple documents) into a number of clusters such that sentences within each cluster are similar. Then we create the mean vector for each cluster. Similar to the procedure in Fig. 9.3, one or more sentences can be selected that are closest to the mean vector from each cluster as its representative sentences. The selected representative sentences from the clusters are displayed as the desired summarization.

In addition to the methods above, topic sentences can also be selected based on some linguistic heuristics. A number of ideas have been proposed in the literature. For example, it has been noticed that sentences that occur in a title, or either very early or very late in a document and its paragraphs, tend to carry much more information about the main topics. Therefore we can select sentences simply based on their positions in an article. Another idea is to use linguistic cue phrases such as *in conclusion*, *the most important*, or *the purpose of the article*, which often indicate important topic sentences.

Although linguistic heuristics such as those described above are helpful for sentence extraction, it is often difficult to determine the relative importance of sentences that are selected based on different heuristics. In order to facilitate a consistent ranking of sentences, a recent trend in topic sentence extraction is to employ machine-learning methods. For example, trainable classifiers have been used to rank sentences based on features such as cue phrase, location, sentence length, word frequency and title, etc. Given a document, we may select a pool of top-ranked sentences based on outputs from the resulting classifier. In order to obtain a more concise summary, an algorithm such as the one in Fig. 9.3 can be employed to eliminate redundant sentences from the pool.

9.2 Active Learning

Prediction methods work with labeled data. We characterized clustering as a way to assign class labels to unlabeled sample data. Prediction methods work best with well-formed goals, and that implies careful problem design with labels clearly expressed and naturally assigned. You define the problem and the computer solves it. Otherwise, you may be asking the computer to define the problem, too.

Another perspective is to look at labels as an expense. We can exert an effort to obtain them, but we need not lead a spendthrift lifestyle. If we need labels, we will "purchase" them. If we have a predictor that is performing perfectly, why go out and get additional labeled data?

Active learning attempts to reduce the number of labeled documents needed for training. The assumption is that labels are expensive, and documents will arrive unlabeled. We can ask someone to label the incoming document, but that will entail an expense. Figure 9.4 illustrates this view of labeling and its effect on the learning process.

How is the decision made to request a label? That is easy for prediction methods that assign a probability to their decisions. They will make a request for a label when they are unsure about their decision. If they have high confidence, they discard the new document. The process does not have to take place in real time. One can collect a number of documents that cannot be classified with confidence, and a request is later made to label them. The decision to request a label is made as follows for a new unlabeled document:

1. Classify the document using the current decision model. Let C be the assigned category.

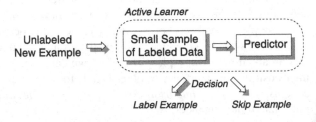

Fig. 9.4 Active learning and labeling

2. If $Pr(C) <$ threshold, request a label; otherwise, ignore the document.

The procedure above can be applied iteratively. Instead of using a predetermined threshold, in practice one can set it so that a small number of least confidently predicted documents are returned. After adding these newly labeled data into the training set, we retrain the classifier. We can then repeat the procedure until a desired accuracy is attained. One potential issue with this method is that similar data are likely to be selected (or ignored) simultaneously. The problem can be alleviated by randomizing the data selection process. For example, at each iteration, we may use a relatively small pool of randomly selected unlabeled data.

Even if the learning algorithm does not assign probability estimates to its decisions, the active-learning procedure can still be applied as long as we can estimate how confident its decisions are. For example, if the decision of a learning method is based on voting many classifiers, then the degree of agreement among the classifiers can be regarded as a measure of confidence.

Experimenting with text categorization and with some numerical data-mining applications, researchers have found that active learning can often drastically reduce their overall sample size while achieving results comparable to training on the full set of labeled documents.

9.3 Learning with Unlabeled Data

Unlabeled documents can be useful for prediction even when labels will never be assigned. For example, global dictionary compilations and tf-idf computations do not need labeled data. Some prediction methods may use pooled covariance matrices that can be computed without labels. A general idea is to find features in unlabeled data. We may discover patterns from unlabeled data and use such patterns as features input to the learning algorithm. If such patterns exist, then unlabeled data can help.

A more open question is whether the assignment of labels can be simulated by programs to improve predictive results. One proposal, called *cotraining*, is based on learning and applying two different classifiers (or one classifier on two independent sets of features) and computing their confidence measures. The procedure is related to active learning. However, at each iteration, instead of choosing the most uncertain data to be labeled by a human, one selects data that can be labeled reliably by one of the classifiers (according to the confidence measures) and simply use the corresponding labels in the next iteration as if they were the true labels. This method only works if we can indeed bootstrap classifiers this way without human intervention. The theoretical analysis of this method is based on the assumption that labels guessed by the algorithm are correct. However, the potential danger is that more and more incorrectly labeled data are included in the training data, which reduce the quality of the underlying classifiers. Promising results have been reported for some problems with small numbers of labeled data to start with. These experiments were performed with labeled data and while hiding some of the labels.

Some researchers have also reported improved results by using the following sequence of steps:

1. Train from labeled data.
2. Apply the resulting classifier to unlabeled data.
3. Add the unlabeled data and the program-assigned labels to the training sample
   and retrain.

This method also has the danger of training a classifier based on incorrectly labeled
data. Although some experiments show promising results, the circumstances of improved results are not yet fully understood.

9.4 Different Ways of Collecting Samples

Predictive methods are implemented in programs that process examples of prior
experience. The examples are assembled into a single sample and processed in one
big batch. Let's examine alternative ways of collecting samples and processing data
in smaller pieces.

9.4.1 Ensembles and Voting Methods

Instead of being combined in one sample, examples could be collected from different samples. We have already noted that the fundamental approach to sampling
for prediction assumes that documents are *i.i.d.*, independent and identically distributed. We know documents change over time, but within a narrow window, the
population may be relatively stationary. We might take a sample periodically, and
instead of retraining a classifier, we can induce a new one from the new sample.
Thus we would have a different classifier for each sample, and we expect that the
samples are from the same population. This simplifies our learning task, but how do
we make decisions from so many classifiers? We can apply each of the classifiers
to a new document and then tabulate their votes. Figure 9.5 illustrates the overall
process. In the simplest form, all votes are equal, and the class with the most votes
wins the election. More complex versions may use weighted voting.

Numerous sampling schemes have been developed to simulate multiple samples
from just a single sample. This is done by resampling or weighting the examples in
the training data. Some techniques, such as *bagging*, randomly draw examples with
replacement (i.e., repeats) from a single sample. Another technique, *boosting*, is a
sampling approach with a memory. It samples the examples that are misclassified
with a higher weight than the correctly classified examples. All these variations
result in multiple classifiers, which are voted, either equally or by weighted voting.

The results for these methods on benchmark data are among the best. Yet, for
practical applications, these methods are less interesting than alternatives. The linear
methods described in Chap. 3 also achieve excellent results with a single straightforward solution. The voting methods can take a long time to train and apply, and their
results are not easy to understand. On the positive side, they may find classifiers

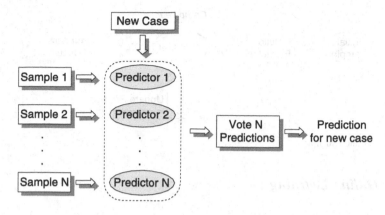

Fig. 9.5 Voting predictors from multiple samples

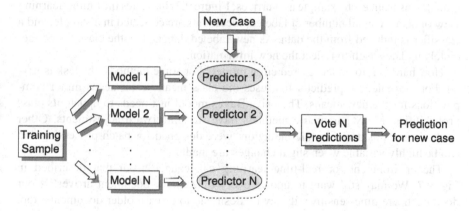

Fig. 9.6 Voting predictors from multiple models

that cover the rare instances that are overlooked by single-sample methods. Reading and comprehending text is a task we all perform. We are reluctant to transform this effort into something that is less well-understood, especially when other approaches are equal to the task.

An alternative to training on multiple samples using the same learning method is to train diverse models and then combine their predictions. This stacking process is shown in Fig. 9.6. For example, one could combine the results of linear models, trees and nearest neighbors. This approach typically involves relatively few models and could be considered a form of mechanical collaboration. Different variations on this theme of combining diverse sources of opinion have proven effective. For example, several teams may work on the same problem using different approaches, and the best results may be achieved when their independent predictions are later combined. Predictions from alternative models may be combined by linear or voting methods, or by tuning and empirical testing.

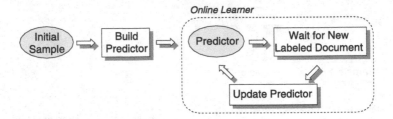

Fig. 9.7 Online learning

9.4.2 Online Learning

A sample consists of one or more (usually many) examples. For real-time learning, we consider a more elemental view of data. Instead of learning from a sample, a program learns from each example as it arrives. Figure 9.7 illustrates the online-learning view of data. A small number of labeled documents are collected in a sample, and a classifier is induced from the data. As new labeled data arrive, the classifier immediately updates itself to reflect the new information.

How hard is it to update a prediction model? For some methods, the task is trivial. For example, the predictor that uses the class mean vector needs minor computations for the new means. The naive Bayes model just needs to update its class probabilities. Many parametric methods have straightforward adjustments. Other methods, such as decision rule induction, were designed for batch processing and can be highly variable when slight changes are made.

The environment for real-time learning can change from that described in Fig. 9.7. We may still want to update every time a new document arrives. If our documents are time-sensitive, it may be desirable to discard older documents, too. The situation may share many of the characteristics of a time series, where we process documents arriving within a fixed window of time.

Online-learning methods can also be used to train linear classifiers. As we have mentioned before, linear classifiers are well-suited for text-mining applications since they can take advantage of large numbers of features. A popular online method for linear classification is the simple and elegant perceptron algorithm. This method is mistake-driven in that it updates the weight vector when the prediction with the current weight makes a mistake.

Using the notation in Chap. 3, we assume that the weight vector is w and the i-th training data are (x^i, y^i), where x^i is the feature vector and y^i is a binary label with value ± 1. We initialize the algorithm with $w = 0$. Then it goes through the data from $i = 1, 2, \ldots$. After examining the i-th data, the algorithm does nothing if $w \cdot x^i y^i > 0$ (which means that it has made a correct prediction). However, if $w \cdot x^i y^i \leq 0$ (that is, the current weight vector has made a mistake on the i-th data), then it updates the weight vector w simply as $w \to w + x^i y^i$.

One can show that if the training data are linearly separable (that is, there exists a weight vector w_* such that $w_* \cdot x^i y^i > 0$ for all i), then by going through the training data repeatedly, the perceptron algorithm will find a weight vector that separates the data in a finite number of steps.

Despite its simplicity, the perceptron algorithm works quite well for many problems. However, its performance is usually lower than more sophisticated linear scoring methods such as the one we described in Chap. 3. Practitioners also find that instead of using the final weight vector when the algorithm is stopped, one often achieves better performance by averaging weight vectors obtained during all steps of the algorithm.

9.4.3 Cost-Sensitive Learning

For binary classification, two types of errors occur: false positives and false negatives. Some learning methods treat these errors equally and try to minimize the overall error rate. Text-mining applications are very sensitive to the tradeoff of these two errors relative to predicting the positive class. Measures of precision and recall capture the distinction between the two types of errors. Typically, a minimum level of precision is required. Increasing one of these measures comes at the expense of the other. The natural way to move the direction of these measures is to assign a fixed cost to errors. Increasing the cost of error for missing a document belonging to the positive class will increase recall. Similarly, increasing the cost of an error for a negative-class document may increase precision. Increasing the cost of error for missing positive documents is equivalent to increasing the number of examples for that class. Most text-mining programs have some variable that can be varied to produce different levels of precision and recall. It's up to the application expert to decide whether the overall results are satisfactory and to find the best compromise in precision and recall.

Varying the costs of error can sometimes improve overall predictive accuracy. Learning programs do not employ optimal algorithms, and tuning learning parameters can have a beneficial effect. Increasing the implied number of examples for the positive class by increasing the cost of errors of omission might force the learning program to be more ambitious in pursuing rules for its class. For the Reuters benchmark, our rule induction program has better accuracy by about 1% when the initial cost of error is doubled for all indexed topics.

Almost all applications of predictive text-mining methods have been for classification. One can readily envision applications where the prediction is measured in numerical terms such as gains or losses. For example, one could attempt to predict the net gain in stock prices based on sampling newswires and other documents. Predicting the gain or loss for an individual stock is a regression application. A sample of newswires for companies of interest would be collected along with the change in their stock price for some fixed period following the release of the newswire. Using text combined with a time-sequenced measurement, a learning method would try to find patterns that identify an upcoming change in a stock price.

Measures more complex than cost of errors might be used. Not only might errors be penalized, but positive decisions might be reinforced. We could assign gains and losses to each decision. The actual gain or loss depends on whether the decision is correct, and the magnitude of the profit or loss is not necessarily uniform for each example.

9.4.4 Unbalanced Samples and Rare Events

Many learning methods tend to be timid about learning rare events. In that situation, the positive class is overwhelmed by the sheer number of negative events. Unless the two classes can be readily distinguished by one or two key words, the method may give up and determine that it's best not to predict that class. If this occurs, one should try to rebalance the class examples for training. A typical rebalance is an equal number of examples in each class, where all positive class examples are used along with a random selection of the negative examples. Testing can be done on new cases without any correction for rare events. This rebalancing has much in common with the cost adjustments described in the previous section. It may be important to detect rare events, even at the expense of increased error on the negative side. Moreover, given the suboptimality of learning methods, increasing the prevalence of the positive cases can jumpstart the method into looking more extensively for solutions.

9.5 Distributed Text Mining

The Internet provides a vast repository of textual data that is impossible to store and process on a single computer. Even privately held corpora are steadily increasing in volume and are often stored in a distributed manner due to practical and privacy considerations. Text mining tools are being increasingly deployed on such data for time-critical goals—not only must they be able to process the massive amounts of text, they must do so in a reasonable amount of time. The approaches of Sect. 9.4 arose out of the need to apply existing algorithms in such situations. An alternative emerging trend is to take advantage of affordable high-performance computing systems to process data using distributed methods. The challenge is to develop new algorithms that can be implemented using parallel/distributed computing infrastructure.

A popular architecture that was originally developed to handle text processing tasks in search engines is based on the *map-reduce* paradigm. The architecture hides the distributed system details from the programmer and handles all synchronization, parallelization, fault-recovery and I/O. As a result, the programmer can concentrate on solving the text-processing task. A key feature of this model is that the programs are written using *map* and *reduce* primitives derived from functional programming languages. Many text mining tasks can be easily recast as a sequence of *map* and *reduce* steps. The popularity of this architecture is partly due to the fact that it relies

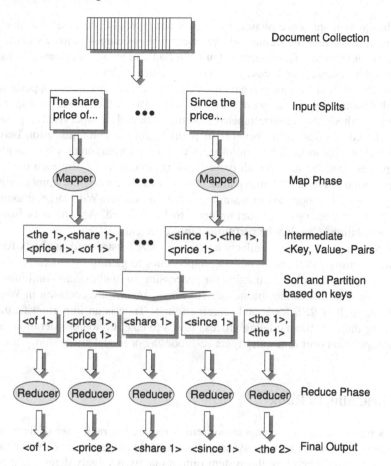

Fig. 9.8 Computing word counts with MapReduce

on cheap commodity hardware. Another reason for the popularity is that a free, high-quality, open-source implementation, called *hadoop*, is available.

Figure 9.8 shows how a list of words and their frequencies can be obtained from a document collection on hadoop. The programmer specifies only the code for the map and reduce steps. The map/reduce phases run independently on different splits of the data. Hadoop takes care of splitting/partitioning the data, assigning processors for the map/reduce steps, synchronizing the phases, etc. The map phase outputs intermediate key-value pairs with the key being a word and the value being 1, to indicate presence of a word. Hadoop then sorts and partitions the pairs by the keys and sends each reducer pairs having the same key. The reducers aggregate the key values to obtain the counts. Clearly, not all tasks can be achieved in a single map-reduce step. A key assumption made in map-reduce programming is that disk space is cheap and limitless and programs can be made to run faster simply by adding more processor nodes. Hence there is no need for any individual map-reduce phase to be very complex. They can simply write out intermediate output for further processing by subsequent map/reduce phases. Input files are typically organized for quick sequential reads. The entire framework operates on large computer clusters made with standard components. The system is fault-tolerant and able to transparently handle the available resources and reassign work from failed nodes.

Distributed Text Mining presents a unique set of challenges and opportunities. With the availability of vast amounts of data, it becomes difficult to apply supervised learning methods that require labeled training data and unsupervised approaches are preferred. To realize the benefits of parallelism, the communication between individual processors must be minimized. Common information may be provided to each processor at the start, but all preprocessing is done locally by each processor. The preferred solution in a distributed data setting is a weighted combination of multiple models as opposed to a single monolithic model. With large amounts of data, basic assumptions in modeling need to be revisited. Are the data from the same population? Can they be partitioned into different vector spaces?

As real-world document collections grow in size, distributed methods for text mining are among the most promising approaches to getting results in reasonable time. They are particularly attractive for processing *streaming data*—unlimited sequences of data that cannot be stored. Text data streams are common in Web applications such as RSS feeds and search-engines. Time is an important feature of streaming data and the modeling process needs to take it into consideration. Shifts in the population over time can trigger re-modeling or model adaptation.

9.6 Learning to Rank

In ranking applications, a computer system is required to rank a set of items based on a given input. Moreover, the system often needs to present only a few top ranked items to the user. Therefore the system output quality is largely determined by the performance near the top of its rank-list.

This problem has become increasingly important. An example is Internet search, where the user presents a query to the search engine, and the search engine then selects a few web-pages that are most relevant to the query from the whole web. The quality of a search engine is largely determined by the top-ranked results the search engine can display on the first page. Therefore the evaluation of ranking systems focus on the top positions.

In practical applications, each available position i can be associated with a weight c_i that measures the importance of that position. Let y_i measure the quality of the result at position i (for example, in web search, if page i is relevant, we may set $y_i = 1$; otherwise, we set $y_i = 0$), the ranking quality is the weighted sum: $\sum_{i=1}^{m} c_i y_i$. This metric is referred to as DCG (discounted cumulated gain).

Much research in ranking has focused on pair-wise preference learning. In this learning model, we are presented with two items x and x', and whether x ranks higher than x' (label is 1); or x ranks lower than x' (label is -1). The goal is to learn a scoring function $f(x)$ such that $f(x) > f(x')$ if x is ranked higher than x'. If the scoring function is linear: $f(x) = w \cdot x + b$, then $f(x) > f(x')$ is equivalent to $w \cdot (x - x') > 0$. Therefore for linear scoring models, we can turn a pair-wise preference learning problem into binary classification: for each pair (x, x'), we create a training datum with input $x - x'$, and the corresponding output label is either 1 (if x ranks higher) or -1 (if x ranks lower). The scoring function $f(\cdot)$ can then be learned with a linear classifier. For nonlinear models, the learning problem is different from binary classification.

The information retrieval problem, which is discussed in Chap. 4, can be considered as a ranking problem. A commercial ranking system such as a web search engine also needs to aggregate information from many different sources, such as tf-idf text matching score, page quality (e.g., page rank score), the popularity of a page (how many times it is clicked), etc. The system takes such information for each item to be ranked (such as a web-page) as its input and produces an aggregated score as its output. Items (such as web-pages) with highest scores are then presented to the user. Therefore how to effectively learn the final scoring function is a fundamental research problem.

9.7 Question Answering

When the Web began to grow rapidly and the number of users who wanted to search the Web also grew rapidly, interest in simple means of query, such as natural language (NL) query, grew also. Most systems that billed themselves as NL systems were actually key word systems in disguise. They stripped the NL input of function words (such as "the" and the like) and ran a conventional search using the remaining content words.

One of the more popular NL query systems was Ask Jeeves, a company that has survived the Internet business collapse. According to comments in response to a patent infringement suit brought by two MIT professors, the system worked in the following way. First, the question is tokenized to find key terms, and then the question is parsed for word meaning by semantic and syntactic networks that were built by Ask Jeeves. After the question is tokenized, it is reorganized into a structure compatible with question templates developed by Ask Jeeves. At this point, the service provides the user with a set of potential locations for the answer to the question.

From the Ask Jeeves example, we can see that the goal of question answering is to find the answer from a collection of text documents to questions in natural language format. Current systems can answer questions such as "In what country did the game of croquet originate?" or "What was the length of the Wright brothers' first flight?" The answer to such questions can be a short passage of text or a sentence that contains the answer. The potential answer phrases can also be marked up in the presentation of the results.

A typical question-answering system, illustrated in Fig. 9.9, contains the following comments:

Fig. 9.9 Question answering
system

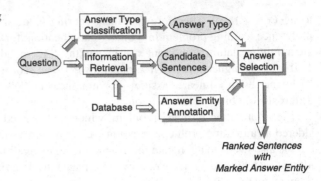

Ranked Sentences
with
Marked Answer Entity

- answer type classification
- answer entity annotation
- information retrieval
- answer selection.

   The answer type classification module matches the incoming question against a
set of predefined templates. These templates determine what kind of information
we are looking for. Accordingly, the question will be assigned a set of potential
answer types such as length, location, etc. Meanwhile, the underlying document
collection is pre-annotated with the answer types using a named entity tagger. The
annotated sentences will be matched against the expected answer type in the answer
selection module to rank and narrow down the potential choices. The information
retrieval (IR) module selects a set of sentences that potentially contain the answer
from the underlying document collection. This is done by using the search technology. However, since the same question can often be asked in different ways, the IR
module includes a query expansion stage, which maps key words in the question
into synonyms or equivalent linguistic forms. The answer selection module ranks
candidates found by the IR module and marks the desired answer entities. In order
to determine whether a sentence or text passage is the desired answer to the question, one may take various heuristic information. Either a hand-crafted system or a
machine-learning system can be used for this purpose.

9.8 Summary

In this chapter, we briefly touch on topics that may increase in importance for text
mining, but are not yet central to prediction. These include summarization, active
learning, learning with unlabeled data, learning with multiple samples or models,
online learning, cost-sensitive learning, unbalanced samples and rare events, distributed text mining, rank learning and question answering.

9.9 Historical and Bibliographical Remarks

Text summarization has been a topic of interest since the early days of text processing under the name of "automatic abstracting" (Luhn 1958). As the amount of online text has continued to grow, the activity has become more organized. In May 1998, the US government completed the TIPSTER Text Summarization Evaluation (SUMMAC), which was the first large-scale, developer-independent evaluation of automatic text summarization systems. SUMMAC, it was claimed, established definitively in a large-scale evaluation that automatic text summarization is very effective in relevance assessment tasks. An overview of the present state of the art can be found in the papers in Radev and Tenfel (2003). Multidocument summarization is also discussed in Radev et al. (2002).

Due to the difficulty of obtaining a large amount of labeled data in many applications, there has been increasing interest in learning with unlabeled data. Active learning is related to an old idea in statistics called sequential design. It has been studied in the machine-learning community both theoretically and empirically. The uncertainty-sampling method appeared in Lewis and Catlett (1994). Approaches that combine active learning with boosting and voting classifiers are given in Liere and Tadepalli (1997) and Iyengar et al. (2000). Many empirical studies suggested that active learning improves the learning performance, but the method still requires significant human efforts in data labeling. It only reduces the amount of human labor by focusing on the most informative samples. Ideally, one hopes to take advantage of unlabeled data without any additional human effort. A number of proposals appeared (Nigam 2001). The idea of cotraining was proposed in Blum and Mitchell (1998). Subsequent experiments suggested that the method works for some problems but not always. As argued in Zhang and Oles (2000), the usefulness of unlabeled data depends on the underlying probability model that generates the data. For certain problems, procedures that simulate the unknown labels can be harmful due to the potential errors introduced into the training data.

Boosting techniques for text categorization are discussed in Schapire and Singer (2000). A variation of voted multiple decision trees was used in Weiss et al. (1999) to produce one of the best reported results for text categorization on the standard Reuters benchmark data. Discussions of stacking techniques for combining predictions from multiple models are found in Bao et al. (2009) and Dzeroski and Ženko (2004). The Netflix contest, a prediction competition with a million dollar prize and thousands of entrants, was won by a collaboration of teams using a stacking approach to learning (Bell et al. 2009).

The perceptron method was first analyzed by Rosenblatt (1962). It was shown that the algorithm makes no more than a certain number of mistakes when the training data are linearly separable. This result has motivated many recent developments in the computational learning theory community (COLT), where other online algorithms with provable mistake bounds were proposed. A particularly interesting algorithm is the Winnow method proposed by Littlestone (1988). Instead of the additive update rule in the perceptron algorithm, the Winnow method uses a multiplicative update scheme. Despite its simplicity, the perceptron method can be quite effective. It has been applied to many applications, including some natural language processing problems (Collins 2002).

The MapReduce model (Dean and Ghemawat 2008) is derived from functional programming and earlier languages such as Lisp. Hadoop (White 2009) began as an open-source reimplementation of Google's proprietary file system (Ghemawat *et al.* 2003). It is now a part of Apache with several subprojects. See http://hadoop.apache.org. Besides Hadoop, there are many other commercial and free implementations of MapReduce and a partial listing can be found in its Wikipedia page.

The pair-wise preference learning model for ranking has been studied by many authors. For example, an SVM based ranking method was considered in Herbrich *et al.* (2000), and a similar method based on AdaBoost was proposed in Freund *et al.* (2003). Machine learning ranking methods were also used in optimizing commercial web-search systems (Burges *et al.* 2005; Cossock and Zhang 2008). The DCG criterion for measuring ranking performance was proposed in Jarvelin and Kekalainen (2000). It allows quality grades for returned items y_i to have multiple values, which generalizes more traditional information retrieval metrics that allow only binary quality grades (either relevant or not relevant).

During the 1970s and early 1980s, natural language question answering was an active research area. Questions were directed at formatted databases rather than text archives. Much of the research was devoted to building systems that could be customized by a user to a new database (Damerau 1985). Unfortunately, the problem proved to be extremely difficult to solve, and researchers moved to other topics. However, the area revived in the late 1990s due to its potential applications in Web searching. Beginning in 1999, the National Institute of Standards and Technology (NIST), which had been sponsoring a Text Retrieval Conference (TREC), added a track for question answering.

In the first conferences, participants returned either 50- or 250-word passages that supposedly contained the answer. In some scientific conferences, participants return only a small snippet of text as the answer to a question, along with a confidence rating for the answer. In 2002, the best system achieved a recall of 80% with a precision of 59%. Two other systems had even better recall but with precision in the low 20% range (Voorhees and Buckland 2002). In 2003, NIST added an additional track, High Accuracy Retrieval from Documents (HARD), in which participants could leverage additional information about the searcher or the search context through techniques such as passage retrieval and using very targeted interaction with the searcher. Details of individual systems can be found in the proceedings of the TREC conferences, available on the NIST Web site.

The Ask Jeeves techniques for question answering are described in Seshasai (2000).

9.10 Questions and Exercises

1. In generating summaries, how can one ensure that they are grammatically correct?

2. Suppose we have a learning method that only learns hard classification rules. How can it be used in an active learning environment?
3. Given a set of documents, describe how document vectors (using the 50 most frequent words as features) can be produced using map/reduce.
4. Discuss the role of linguistic information in question-answering systems.
5. Consider a system such as *Yahoo! Answers*. Discuss how it might be designed and what are some of the problems with evaluating it.

Appendix A
Software Notes

We have examined a number of different techniques for text mining, and some methods have been described in algorithmic form. Undoubtedly, readers will want to try these methods to gain firsthand experience with them. Others might be interested in applying the techniques to their own data. In the preceding chapters, references have been made to software implementations accompanying this book.

The software is provided by Data-Miner Pty. Ltd. A free single-user license is included for those who have purchased the book. These notes include a brief description of the software, provide details of the hardware/software requirements, and give instructions on how to get the software. Full details are available online at Data-Miner's Web site (http://www.data-miner.com).

A.1 Summary of Software

The software is provided in three parts. The first part is the text-mining software kit (TMSK), which includes routines for preprocessing XML-based text documents. The second part is a complete package for rule-based text classification (RIKTEXT). Both RIKTEXT and TMSK share the same data format for vectors. TMSK can be used to prepare data for RIKTEXT. The third part is a set of linux scripts and sample data files. The data files are used in the exercises of many chapters.

Table A.1 summarizes the tasks accomplished by the software. The table also lists the section of the book where the underlying algorithms are discussed. All the key tasks discussed in the book are covered. Sample data are provided and the documentation describes how users may specify the XML format of their own input documents. The documentation also lists existing sources of XML documents compatible with the software.

Except for rule-based classification, which is implemented as RIKTEXT, everything else is included in TMSK or in the Linux-based scripts. RIKTEXT complements TMSK by providing methods for constructing and using rules for document classification. The input data format for RIKTEXT is identical to that of the classification methods in TMSK. However, RIKTEXT has a number of options that are especially helpful for rule-based systems.

S.M. Weiss et al., *Fundamentals of Predictive Text Mining*,
Texts in Computer Science 41,
DOI 10.1007/978-1-84996-226-1, © Springer-Verlag London Limited 2010

Table A.1 Tasks accomplished by the accompanying software

Text to vectors	Tokenization	Sect. 2.3
	Stemming	Sect. 2.4
	Dictionary creation	Sect. 2.5
	Vector generation	Sect. 2.5
	End-of-Sentence detection	Sect. 2.6
Prediction	Rule-based classifiers	Sect. 3.4.3
	Naive Bayes	Sect. 3.4.4
	Linear models	Sect. 3.4.5
Information retrieval	Document/query matcher	Sect. 4.7
Finding structure	k-means clustering	Sect. 5.2.1
Information extraction	Named entity identification	Sect. 6.2.2
Data sourcing	Linux-based scripts	Chap. 7

A.2 Requirements

Hardware

- TMSK is available for any hardware that has a Java interpreter.
- RIKTEXT is provided for Intel-compatible (x86) machines.

Software

- TMSK uses Java version 1.3.1 or higher. Its modules are run as Java applications with command-line arguments.
- RIKTEXT runs on Linux or Windows. Most modern Linux installations are compatible with RIKTEXT. On Microsoft Windows, RIKTEXT is run in a Command-Prompt or MS-DOS window.
- The Linux scripts run on most modern linux distributions.

A.3 Download Instructions

A free single-user license is provided to those who have purchased the book. The software is provided by Data-Miner Pty. Ltd. and can be downloaded from their Web site. Visit http://www.data-miner.com and follow the links for TMSK and RIKTEXT. The single-user license can be viewed there, and users must accept its terms and declare that they have purchased the book prior to download. Documentation for the software is available online and can be reviewed prior to download as well. After accepting the terms of the software, a user name and password must be provided to get the software. The user name is `tmskriktext` and the password is the 12 digit number 780387954332. The user name is case-sensitive. After downloading the software, follow the installation instructions provided online.

For the scripts and exercises, additional files are needed. See the download section of http://www.data-miner.com for the link labeled *Files for Exercises in Text-Mining Book*. Download and unzip the file ExerciseFiles.zip and note the five files in it: scripts.zip, stemwds.txt, stopword.txt, trn.zip and tst.zip.

References

C. Apté, F. Damerau, and S. Weiss. Automated learning of decision rules for text categorization. *ACM Transactions on Information Systems*, 12(3):233–251, 1994.

A. Banerjee and J. Langford. An objective evaluation criterion for clustering. In *Proceedings of KDD-2004*. ACM, New York, 2004.

X. Bao, L. Bergman, and R. Thompson. Stacking recommendation engines with additional meta-features. In *RecSys'09: Proceedings of the Third ACM Conference on Recommender Systems*, pages 109–116. ACM, New York, 2009.

R. Bekkerman, R. El-Yaniv, N. Tishby, and Y. Winter. Distributional word clusters vs. words for text categorization. *Journal of Machine Learning Research*, 3:1183–1208, 2003.

R. Bell, J. Bennett, Y. Koren, and C. Volinsky. The million dollar programming prize. *IEEE Spectrum*, pages 28–33, 2009.

D. Bikel, S. Miller, R. Schwartz, and R. Weischedel. Nymble: A high-performance learning name finder. In *The Fifth Conference on Applied Natural Language Processing*, pages 194–201. ACM, New York, 1997.

D. Bikel, R. Schwartz, and R. Weischedel. An algorithm that learns what's in a name. *Machine Learning*, 34(1–3):211–231, 1999.

A. Blum and T. Mitchell. Combining labeled and unlabeled data with co-training. In *Proceedings of the Eleventh Annual Conference on Computational Learning Theory*, pages 92–100. ACM, New York, 1998.

A. Borthwick. *A maximum entropy approach to named entity recognition*. Ph.D. thesis, New York University, 1999.

L. Breiman. Prediction games and arcing algorithms. *Neural Computation*, 11:1493–1517, 1999.

E. Brill. Transformation-based error-driven learning and natural language processing: A case study in part-of-speech tagging. *Computational Linguistics*, 21(4):543–565, 1995. http://www.cis.upenn.edu/~adwait/penntools.html.

C. Burges, T. Shaked, E. Renshaw, A. Lazier, M. Deeds, N. Hamilton, and G. Hullender. Learning to rank using gradient descent. In *ICML'05*, 2005.

M. Califf and R. Mooney. Relational learning of pattern-match rules for information extraction. In *Working Notes of AAAI Spring Symposium on Applying Machine Learning to Discourse Processing*, pages 6–11. AAAI Press, Menlo Park, 1998.

C. Cardie and K. Wagstaff. Noun phrase coreference as clustering. In *Proceedings of the Joint SIGDAT Conference on Empirical Methods in NLP and Very Large Corpora*, pages 82–89. ACL, East Stroudsburg, 1999.

E. Charniak. Statistical techniques for natural language parsing. *AI Magazine*, 18(4):33–43, 1997.

D. Chiang. Statistical parsing with an automatically-extracted tree adjoining grammar. In *Proceedings of the ACL 2000*, pages 456–463. ACL, East Stroudsburg, 2000.

W. Cohen. Learning rules that classify email. In *Proceedings of the AAAI Spring Symposium on Machine Learning in Information Access*. AAAI Press, Menlo Park, 1996.

S.M. Weiss et al., *Fundamentals of Predictive Text Mining*,
Texts in Computer Science 41,
DOI 10.1007/978-1-84996-226-1, © Springer-Verlag London Limited 2010

M. Collins. Discriminative training methods for hidden Markov models: Theory and experiments with perceptron algorithms. In *Proceedings of EMNLP'02*. ACL, East Stroudsburg, 2002.

D. Cossock and T. Zhang. Statistical analysis of Bayes optimal subset ranking. *IEEE Transactions on Information Theory*, 54(11):5140–5154, 2008.

M. Craven and S. Slattery. Relational learning with statistical predicate invention: Better models for hypertext. *Machine Learning*, 43:97–119, 2001.

D. Cutting, D. Karger, J. Pedersen, and J. Tukey. Scatter/Gather: A cluster-based approach to browsing large document collections. In *Proceedings of SIGIR-92*, pages 1–12. ACM, New York, 1992.

R. D'Agostino, R. Vasan, M. Pencina, P. Wolf, M. Cobain, J. Massaro, and W. Kannel. General cardiovascular risk profile for use in primary care: the framingham heart study. *Circulation*, 743–753, 2008. http://www.framinghamheartstudy.org/risk/gencardio.html.

F. Damerau. Problems and some solutions in customization of natural language database front ends. *ACM Transactions on Information Systems*, 3(2):165–184, 1985.

F. Damerau, T. Zhang, S. Weiss, and N. Indurkhya. Text categorization for a comprehensive time-dependent benchmark. *Information Processing and Management*, 40(2):209–221, 2004.

J. Darroch and D. Ratcliff. Generalized iterative scaling for log-linear models. *The Annals of Mathematical Statistics*, 43:1470–1480, 1972.

J. Dean and S. Ghemawat. Mapreduce: Simplified data processing on large clusters. *Communications of the ACM*, 51(1):107–113, 2008.

A. Dempster, N. Laird, and D. Rubin. Maximum likelihood from incomplete data via the EM algorithm. *Journal of the Royal Statistical Society Series B*, 39(1):1–38, 1977. With discussion.

I. Dhillon and D. Modha. Concept decompositions for large sparse text data using clustering. *Machine Learning*, 42(1):143–175, 2001.

S. Dumais and H. Chen. Hierarchical classification of web content. In *Proceedings of the 23rd ACM International Conference on Research and Development in Information Retrieval*, pages 256–263. ACM, New York, 2000.

S. Dzeroski and B. Ženko. Is combining classifiers with stacking better than selecting the best one? *Machine Learning*, 54(3):255–273, 2004.

J. Earley. An efficient context-free parsing algorithm. *Communications of the ACM*, 13(2):94–102, 1970.

C. Eastman and S. Weiss. A tree algorithm for nearest neighbor searching in document retrieval systems. In *Proceedings of the ACM-SIGIR International Conference on Information Storage and Retrieval*, pages 131–149. ACM, New York, 1978.

C. Elkan. Using the triangle inequality to accelerate k-means. In *Proceedings of the Twentieth International Conference on Machine Learning*, pages 147–153. AAAI Press, Menlo Park, 2003.

C. Feldbaum, editor. *Wordnet: An Electronic Lexical Database*. MIT Press, Cambridge, 1998.

R. Feldman and L. Ungar. Applied text mining, tutorial. In *Proceedings of KDD-2009*. ACM, New York, 2009. http://www.cis.upenn.edu/~ungar/KDD/KDD_tutorial.pdf.

R. Florian, A. Ittycheriah, H. Jing, and T. Zhang. Named entity recognition through classifier combination. In *Proceedings of CoNLL-2003*, pages 168–171. ACL, East Stroudsburg, 2003.

G. Forman. An extensive empirical study of feature selection metrics for text classification. *Journal of Machine Learning Research*, 3:1289–1305, 2003.

D. Freitag. Information extraction from HTML: Application of a general machine learning approach. In *Proceedings of the 15th National Conference on Artificial Intelligence*, pages 517–523. AAAI Press, Menlo Park, 1998.

Y. Freund and R. Schapire. A decision-theoretic generalization of on-line learning and an application to boosting. *Journal of Computer and System Sciences*, 55(1):119–139, 1997.

Y. Freund, R. Iyer, R. Schapire, and Y. Singer. An efficient boosting algorithm for combining preferences. *JMLR*, 4:933–969, 2003.

J. Friedman, T. Hastie, and R. Tibshirani. Additive logistic regression: A statistical view of boosting. *The Annals of Statistics*, 28(2):337–407, 2000. With discussion.

E. Garfield. Citation analysis as a tool in journal evaluation. *Science*, 178:471–479, 1972.

S. Ghemawat, H. Gobioff, and S.-T. Leung. The google file system. *SIGOPS Operating Systems Review*, 37(5):29–43, 2003.

P. Graham. Better Bayesian filtering. In *Proceedings of the 2003 Spam Conference*, 2003. http://spamconference.org/proceedings2003.html.

T. Hastie, R. Tibshirani, and J. Friedman. *The Elements of Statistical Learning: Data Mining, Inference, and Prediction*, Springer Series in Statistics. Springer, New York, 2001.

P. Hayes and S. Weinstein. Construe/tis: A system for content-based indexing of a database of news stories. In *Proceedings of the 2nd Conference on Innovative Applications of Artificial Intelligence*, pages 49–66. AAAI Press, Menlo Park, 1990.

R. Herbrich, T. Graepel, and K. Obermayer. Large margin rank boundaries for ordinal regression. In B. Schölkopf, A. Smola, P. Bartlett and D. Schuurmans, editors, *Advances in Large Margin Classifiers*, pages 115–132. MIT Press, Cambridge, 2000.

M. Hu and B. Liu. Mining and summarizing customer reviews. In *KDD'04: Proceedings of the tenth ACM SIGKDD International Conference on Knowledge Discovery and Data Mining*, pages 168–177. ACM, New York, 2004.

S. Huffman. Learning information extraction patterns from examples. In *IJCAI Workshop on New Approaches to Learning for Natural Language Processing*, pages 246–260. IJCAI, San Francisco, 1995.

N. Ide and J. Véronis. Word sense disambiguation: The state of the art. *Computational Linguistics*, 24(1):1–40, 1998.

N. Indurkhya and F. Damerau, editors. *Handbook of Natural Language Processing, Second Edition*. CRC Press/Taylor and Francis, Boca Raton/London, 2010.

V. Iyengar, C. Apté, and T. Zhang. Active learning using adaptive resampling. In *The Sixth ACM SIGKDD International Conference on Knowledge Discovery and Data Mining*, pages 91–98. ACM, New York, 2000.

N. Jardine and C. van Rijsbergen. The use of hierarchical clustering in information retrieval. *Information Storage and Retrieval*, 7:217–240, 1971.

K. Jarvelin and J. Kekalainen. IR evaluation methods for retrieving highly relevant documents. In *SIGIR'00*, pages 41–48, 2000.

E. Jaynes. Information theory and statistical mechanics. *Physical Review*, 106:620–630, 1957.

T. Joachims. Text categorization with support vector machines: Learning with many relevant features. In *Proceedings of the 10th European Conference on Machine Learning*. Springer, New York, 1998.

D. Jurafsky and J. Martin. *Speech and Language Processing: An Introduction to Natural Language Processing, Computational Linguistics, and Speech Recognition*. Pearson, Upper Saddle River, 2008.

M. Kearns, Y. Mansour, and A.-Y. Ng. An information-theoretic analysis of hard and soft assignment methods for clustering. In *Proceedings of the Thirteenth Conference on Uncertainty in Artificial Intelligence*, pages 282–293. Morgan Kaufmann, San Francisco, 1997.

J. Kim and D. Moldovan. Acquisition of linguistic patterns for knowledge-based information extraction. *IEEE Transactions on Knowledge and Data Engineering*, 7(5):713–724, 1995.

J. Kleinberg. Authoritative sources in a hyperlinked environment. *Journal of the ACM*, 46(5):604–632, 1999.

A. Kolcz and J. Alspector. SVM-based filtering of e-mail spam with content-specific misclassification costs. In *Proceedings of Workshop on Text Mining, IEEE ICDM-2001*. IEEE Press, New York, 2001.

G. Krupka and K. Hausman. IsoQuest Inc.: Description of the NetOwl TM extractor system as used for MUC-7. In *Proceedings of the Seventh Message Understanding Conference (MUC-7)*. NIST, Washington, 1998.

T. Kudoh and Y. Matsumoto. Use of support vector learning for chunk identification. In *Proceedings of CoNLL-2000 and LLL-2000*, pages 142–144. ACL, East Stroudsburg, 2000.

J. Lafferty, A. McCallum, and F. Pereira. Conditional random fields: Probabilistic models for segmenting and labeling sequence data. In *Proceedings of ICML-01*, pages 282–289. Morgan Kaufmann, San Francisco, 2001.

S. Lappin and H. Leass. An algorithm for pronominal anaphora resolution. *Computational Linguistics*, 20(4):535–561, 1994.

J. Leskovec, L. Backstrom, and J. Kleinberg. Meme-tracking and the dynamics of the news cycle. In *Proceedings of KDD-2009*, page 297. ACM, New York, 2009.

D. Lewis. Feature selection and feature extraction for text categorization. In *Proceedings of the Speech and Natural Language Workshop*, pages 212–217. Morgan Kaufmann, San Francisco, 1992.

D. Lewis and J. Catlett. Heterogeneous uncertainty sampling for supervised learning. In *Proceedings of the Eleventh International Conference on Machine Learning*, pages 148–156. Morgan Kaufmann, San Francisco, 1994.

D. Lewis, Y. Yang, T. Rose, and F. Li. RCV1: A new benchmark collection for text categorization research. *Journal of Machine Learning Research*, 5:361–397, 2004.

F. Li and Y. Yang. A loss function analysis for classification methods in text categorization. In *Proceedings of the Twentieth International Conference on Machine Learning*, pages 472–479. AAAI Press, Menlo Park, 2003.

R. Liere and P. Tadepalli. Active learning with committees for text categorization. In *Proceedings of the 14th National Conference on Artificial Intelligence*, pages 591–596. AAAI Press, Menlo Park, 1997.

N. Littlestone. Learning quickly when irrelevant attributes abound: A new linear-threshold algorithm. *Machine Learning*, 2:285–318, 1988.

H. Luhn. The automatic creation of literature abstracts. *IBM Journal of Research and Development*, 2(2):159–165, 1958.

H. Luhn. Auto-encoding of documents for information retrieval systems. In M. Boaz, editor, *Modern Trends in Documentation*, pages 45–58. Pergamon Press, London, 1959.

J. MacQueen. Some methods for classification and analysis of multivariate observations. In *Proceedings of the Fifth Berkeley Symposium on Mathematical Statistics and Probability*, pages 281–297. University of California Press, Berkeley, 1967.

M. Maloof. Incremental rule learning with partial instance memory for changing concepts. In *Proceedings of the International Joint Conference on Neural Networks (IJCNN'03)*, pages 2764–2769. IEEE Press, New York, 2003.

M. Maron and J. Kuhns. On relevance, probabilistic indexing and information retrieval. *Journal of the ACM*, 7:216–244, 1960.

B. Masand, G. Linoff, and D. Waltz. Classifying news stories using memory based reasoning. In *Proceedings of the 15th Annual International ACM SIGIR Conference on Research and Development in Information Retrieval*, pages 59–65. ACM, New York, 1992.

A. McCallum and K. Nigam. A comparison of event models for naive Bayes text classification. In *AAAI/ICML-98 Workshop on Learning for Text Categorization*, pages 41–48. AAAI Press, Menlo Park, 1998.

J. McCarthy and W. Lehnert. Using decision trees for coreference resolution. In *Proceedings of the 14th International Joint Conference on Artificial Intelligence*, pages 1050–1055. Morgan Kaufmann, San Francisco, 1995.

M. McCord. Slot grammar: A system for simple construction of practical natural language grammars. In *Proceedings of the International Symposium on Natural Language and Logic*, pages 118–145. Springer, New York, 1989.

K. McKeown, R. Barzilay, D. Evans, V. Hatzivassiloglou, J. Klavans, A. Nenkova, C. Sable, B. Schiffman, and S. Sigelman. Tracking and summarizing news on a daily basis with Columbia's newsblaster. In *Proceedings of the Human Languages Technology Conference*. ACL, East Stroudsburg, 2002.

P. Melville, W. Gryc, and R. Lawrence. Sentiment analysis of blogs by combining lexical knowledge with text classification. In *Proceedings of KDD-2009*, page 1275. ACM, New York, 2009.

A. Mikheev, C. Grover, and M. Moens. Description of the LTG system used for MUC-7. In *Proceedings of the Seventh Message Understanding Conference (MUC-7)*. NIST, Washington, 1998.

S. Miller, M. Crystal, H. Fox, L. Ramshaw, R. Schwartz, R. Stone, and R. Weischedel. BBN: Description of the SIFT system as used for MUC-7. In *Proceedings of the Seventh Message Understanding Conference (MUC-7)*. NIST, Washington, 1998.

K. Nigam. *Using unlabeled data to improve text classification*. Ph.D. thesis, Carnegie Mellon University, 2001.

K. Nigam, A. McCallum, S. Thrun, and T. Mitchell. Text classification from labeled and unlabeled documents using EM. *Machine Learning*, 39(2/3):1–32, 2000.

L. Page and S. Brin. The anatomy of a search engine. In *Proceedings of the 7th International WWW Conference (WWW 98)*. Brisbane, Australia, 1998. http://www7.scu.edu.au.

L. Page, S. Brin, R. Motwani, and T. Winograd. The PageRank citation ranking: Bringing order to the web. *Stanford Digital Libraries Technologies Project*, 1998.

B. Pang and L. Lee. Opinion mining and sentiment analysis. *Foundations and Trends in Information Retrieval*, 2(1–2):1–135, 2008.

S. Pietra, V. Pietra, and J. Lafferty. Inducing features of random fields. *IEEE Transactions on Pattern Analysis and Machine Intelligence*, 19(4):380–393, 1997.

M. Porter. An algorithm for suffix stripping. *Program*, 14(3):130–137, 1980.

D. Radev and S. Tenfel, editors. *Proceedings of the HLT NAACL 2003 Workshop on Text Summarization*. ACL, East Stroudsburg, 2003.

D. Radev, M. Topper, and A. Winkel. Multi-document centroid-based text summarization. In *Proceedings of ACL-02 Demo Session*, pages 112–113. ACL, East Stroudsburg, 2002.

L. Ramshaw and M. Marcus. Text chunking using transformation-based learning. In *Proceedings of the Third Workshop on Very Large Corpora*, pages 82–94. ACL, East Stroudsburg, 1995.

A. Ratnaparkhi. A maximum entropy part-of-speech tagger. *Computational Linguistics*, 21(4):543–565, 1995. http://www.cis.upenn.edu/~adwait/penntools.html.

A. Ratnaparkhi. Learning to parse natural language with maximum entropy models. *Machine Learning*, 34:151–178, 1999.

E. Ray. *Learning XML*. O'Reilly & Associates, Sebastopol, 2001.

E. Riloff. Automatically constructing a dictionary for information extraction tasks. In *Proceedings of the 11th National Conference on Artificial Intelligence*, pages 811–816. AAAI Press, Menlo Park, 1993.

S. Robertson, S. Walker, S. Jones, M. Hancock-Beaulieu, and M. Gatford. Okapi at TREC-3. In *Proceedings of the Third Text Retrieval Conference*, pages 109–126. NIST, Washington, 1994. http://trec.nist.gov/pubs/trec3/papers/city.ps.gz.

F. Rosenblatt. *Principles of Neurodynamics: Perceptrons and the Theory of Brain Mechanisms*. Spartan, New York, 1962.

D. Roth and W. Yih. Relational learning via propositional algorithms: An information extraction case study. In *Proceedings of the 17th International Joint Conference on Artificial Intelligence*, pages 1257–1263. Morgan Kaufmann, San Francisco, 2001.

H. Rui, A. Whinston, and E. Winkler. Follow the tweets. *Wall Street Journal, Technology section*, 30 November 2009.

G. Salton. A document retrieval system for man-machine interaction. In *Proceedings of the 19th Annual International ACM National Conference*, pages L2.3.1–L2.3.20. ACM, New York, 1964.

G. Salton. *The SMART Retrieval System*. Prentice-Hall, Englewood Cliffs, 1971.

G. Salton and M. Lesk. The SMART automatic document retrieval system—An illustration. *Communications of the ACM*, 8(6):391–398, 1965.

G. Salton and M. Lesk. Computer evaluation of indexing and text processing. *Journal of the Association for Computing Machinery*, 15(1):8–36, 1968.

G. Salton and H. Wu. A term weighting model based on utility theory. In *Proceedings of SIGIR*, pages 9–22. ACM, New York, 1980.

G. Salton, A. Wong, and C. Yang. A vector space model for automatic indexing. *Communications of the ACM*, 18:613–620, 1975.

E. Sang and S. Buchholz. Introduction to the CoNLL-2000 shared task: Chunking. In *Proceedings of the CoNLL-2000 and LLL-2000*, pages 127–132. ACL, East Stroudsburg, 2000.

E. Sang and F. De Meulder. Introduction to the CoNLL-2003 shared task: Language independent named entity recognition. In W. Daelemans and M. Osborne, editors, *Proceedings of CoNLL-2003*, pages 142–147. ACL, East Stroudsburg, 2003.

R. Schapire and Y. Singer. Improved boosting algorithms using confidence-rated predictions. *Machine Learning*, 37:297–336, 1999.

R. Schapire and Y. Singer. BoosTexter: A boosting-based system for text categorization. *Machine Learning*, 39(2/3):135–168, 2000.

S. Seshasai. Winston, Katz sue Ask Jeeves: AI lab researchers attempt to enforce natural language patent. *The Tech (MIT)*, 2000. http://www-tech.mit.edu/V119/N66/.

S. Soderland. Learning information extraction rules for semi-structured and free text. *Machine Learning*, 34(1–3):233–272, 1999.

S. Soderland, D. Fisher, J. Aseltine, and W. Lehnert. CRYSTAL: Inducing a conceptual dictionary. In *Proceedings of the 14th International Joint Conference on Artificial Intelligence*, pages 1314–1319. Morgan Kaufmann, San Francisco, 1995.

W.-M. Soon, H.-T. Ng, and C.-Y. Lim. A machine learning approach to coreference resolution of noun phrases. *Computational Linguistics*, 27(4):521–544, 2001.

G. Stein, A. Bagga, and G. Wise. Multi-document summarization: Methodologies and evaluations. In *Proceedings of the 7th Conference on Automatic Natural Language Processing (TALN'00)*, pages 337–346. ATALA Press, Paris, 2000.

C.-M. Tan, Y.-F. Wang, and C.-D. Lee. The use of bigrams to enhance text categorization. *Information Processing and Management*, 38(4):529–546, 2002.

P. Tan, H. Blau, S. Harp, and R. Goldman. Textual data mining of service center call records. In *Proceedings of KDD-2000*, pages 417–423. ACM, New York, 2000.

B. Taskar, C. Guestrin, and D. Koller. Max-margin Markov networks. In S. Thrun, L. Saul and B. Schölkopf, editors, *Advances in Neural Information Processing Systems 16*. MIT Press, Cambridge, 2004.

C. Tillmann and T. Zhang. An online relevant set algorithm for statistical machine translation. *IEEE Transactions on Audio, Speech, and Language Processing*, 16(7):1274–1286, 2008.

M. Tomita. *Efficient Parsing for Natural Language*. Kluwer Academic, Dordrecht, 1985.

I. Tsochantaridis, T. Joachims, T. Hofmann, and Y. Altun. Large margin methods for structured and interdependent output variables. *JMLR*, 6:1453–1484, 2005.

V. Vapnik. *Statistical Learning Theory*. Wiley, New York, 1998.

E. Voorhees. The cluster hypothesis revisited. In *Proceedings of SIGIR-85*, pages 188–196. ACM, New York, 1985.

E. Voorhees and L. Buckland, editors. *NIST Special Publication 500-251: The Eleventh Text Retrieval Conference (TREC 2002)*, Gaithersburg, Maryland, 19–22 November 2002. NIST Press, Washington, 2002. Co-sponsored by DARPA and ARDA.

D. Walker, D. Clements, M. Darwin, and J. Amtrup. Sentence boundary detection: A comparison of paradigms for improving MT quality. In *Proceedings of the Eighth Machine Translation Summit*. ACL, East Stroudsburg, 2001.

S. Weiss and N. Verma. A system for real-time competitive market intelligence. In *Proceedings of SIGKDD-2002*. ACM, New York, 2002.

S. Weiss, C. Apté, F. Damerau, D. Johnson, F. Oles, T. Goetz, and T. Hampp, Maximizing text-mining performance. *IEEE Intelligent Systems*, 14(4):63–69, 1999.

S. Weiss, B. White, and C. Apté. Lightweight document clustering. In *Proceedings of PKDD-2000*, pages 665–672. Springer, New York, 2000a.

S. Weiss, B. White, C. Apté, and F. Damerau. Lightweight document matching for help-desk applications. *IEEE Intelligent Systems*, 15(2):57–61, 2000b.

T. White. *Hadoop: The Definitive Guide*. O'Reilly Media, Sebastopol, 2009.

P. Willett. Recent trends in hierarchic document clustering. *Information Processing and Management*, 24:577–597, 1988.

J. Xu and B. Croft. Corpus-based stemming using cooccurrence of word variants. *ACM Topics on Information Systems*, 16(1):61–81, 1998.

Y. Yang and J. Pedersen. A comparative study of feature selection in text categorization. In *Proceedings of the Fourteenth International Conference on Machine Learning*, pages 412–420. Morgan Kaufmann, San Francisco, 1997.

D. Zelenko, C. Aone, and A. Richardella. Kernel methods for relation extraction. *Journal of Machine Learning Research*, 3:1083–1106, 2003.

T. Zhang. On the dual formulation of regularized linear systems. *Machine Learning*, 46:91–129, 2002.

T. Zhang. Statistical behavior and consistency of classification methods based on convex risk minimization. *The Annals of Statistics*, 32(1):56–134, 2004. With discussion.

T. Zhang and D. Johnson. A robust risk minimization based named entity recognition system. In *Proceedings of CoNLL-2003*, pages 204–207. ACL, East Stroudsburg, 2003.

T. Zhang and F. Oles. A probability analysis on the value of unlabeled data for classification problems. In *Proceedings of ICML-00*, pages 1191–1198. Morgan Kaufmann, San Francisco, 2000.

T. Zhang and F. Oles. Text categorization based on regularized linear classification methods. *Information Retrieval*, 4:5–31, 2001.

T. Zhang, F. Damerau, and D. Johnson. Text chunking based on a generalization of Winnow. *Journal of Machine Learning Research*, 2(5):615–637, 2002.

T. Zhang, F. Damerau, and D. Johnson. Updating an NLP system to fit new domains: An empirical study on the sentence segmentation problem. In *Proceedings of the Seventh Conference on Natural Language Learning, CoNLL-2003*, pages 56–62. ACL, East Stroudsburg, 2003.

S. Zhong and J. Ghosh. A unified framework for model-based clustering. *Journal of Machine Learning Research*, 4:1001–1037, 2003.

Author Index

S.M. Weiss et al., *Fundamentals of Predictive Text Mining*,
Texts in Computer Science 41,
DOI 10.1007/978-1-84996-226-1, © Springer-Verlag London Limited 2010

Subject Index

A

Accuracy, 10, 67
Active learning, 192
Applications
  classification, 69
  clustering, 107
  criminal justice, 135
  extraction systems, 134
  information extraction, 133
  information retrieval, 133
  intelligence, 135
Ask Jeeves, 201
Ask.com search engine, 177
Attributes, *see* Features

B

Bagging, 194
Bayes rule, 56
Bing search engine, 180
Boosting, 194

C

Case studies
  classification, 157, 169, 174, 181, 184
  clustering, 165, 184
  digital libraries, 161
  document matching, 161
  e-mail filters, 174
  help desk, 165
  market intelligence, 157
  model cases, 165
  named entities, 181
  newspapers, 184
  newswire articles, 169
  search engines, 177

Categorization, *see* Classification
Centroid classifier, 98
Classification, 6, 43
  applications, 69
  case studies, 157, 169, 174, 181, 184
  centroid, 98
  decision rules, 48
  decision trees, 54, 71
  GIS method, 120
  linear, 58
  maximum entropy, 118
  multinomial model, 57
  naive Bayes, 55
  nearest-neighbor, 45
  tag prediction, 117
  voting, 194
Clustering, 7, 91, 189
  applications, 107
  case studies, 165, 184
  descriptors, 105
  hard, *see* k-means
  hierarchical, 99
  k-means, 96
  nearest-neighbor, 93
  soft, 102
Collocations, *see* Features, multiword
Conditional Random Field, 128
Coreference resolution, 129
Corpus
  Reuters, 14, 37
  Wall Street Journal, 31
Cosine similarity, 47, 79
Cost-sensitive learning, 197
Cotraining, 193

S.M. Weiss et al., *Fundamentals of Predictive Text Mining,*
Texts in Computer Science 41,
DOI 10.1007/978-1-84996-226-1, © Springer-Verlag London Limited 2010